FROM ASSOCIATIONS TO RULES

Connectionist Models of Behavior and Cognition

PROGRESS IN NEURAL PROCESSING*

Series Advisor
Alan Murray *(University of Edinburgh)*

Vol. 4: Analogue Imprecision in MLP Training
by Peter J. Edwards & Alan F. Murray

Vol. 5: Applications of Neural Networks in Environment, Energy, and Health
Eds. Paul E. Keller, Sherif Hashem, Lars J. Kangas & Richard T. Kouzes

Vol. 6: Neural Modeling of Brain and Cognitive Disorders
Eds. James A. Reggia, Eytan Ruppin & Rita Sloan Berndt

Vol. 7: Decision Technologies for Financial Engineering
Eds. Andreas S. Weigend, Yaser Abu-Mostafa & A.-Paul N. Refenes

Vol. 8: Neural Networks: Best Practice in Europe
Eds. Bert Kappen & Stan Gielen

Vol. 9: RAM-Based Neural Networks
Ed. James Austin

Vol. 10: Neuromorphic Systems: Engineering Silicon from Neurobiology
Eds. Leslie S. Smith & Alister Hamilton

Vol. 11: Radial Basis Function Neural Networks with Sequential Learning
Eds. N. Sundararajan, P. Saratchandran & Y.-W. Lu

Vol. 12: Disorder Versus Order in Brain Function: Essays in Theoretical Neurobiology
Eds. P. Århem, C. Blomberg & H. Liljenström

Vol. 13: Business Applications of Neural Networks: The State-of-the-Art of Real-World Applications
Eds. Paulo J. G. Lisboa, Bill Edisbury & Alfredo Vellido

Vol. 14: Connectionist Models of Cognition and Perception
Eds. John A. Bullinaria & Will Lowe

Vol. 15: Connectionist Models of Cognition and Perception II
Eds. Howard Bowman & Christophe Labiouse

Vol. 16: Modeling Language, Cognition and Action
Eds. Angelo Cangelosi, Guido Bugmann & Roman Borisyuk

*For the complete list of titles in this series, please write to the Publisher.

Progress in Neural Processing • 17

Proceedings of the Tenth Neural Computation and Psychology Workshop

FROM ASSOCIATIONS TO RULES

Connectionist Models of Behavior and Cognition

Dijon, France 12 – 14 April 2007

Editors

Robert M. French
French National Center for Scientific Research & University of Burgundy, France

Elizabeth Thomas
University of Burgundy, France

World Scientific

NEW JERSEY • LONDON • SINGAPORE • BEIJING • SHANGHAI • HONG KONG • TAIPEI • CHENNAI

Published by

World Scientific Publishing Co. Pte. Ltd.

5 Toh Tuck Link, Singapore 596224

USA office: 27 Warren Street, Suite 401-402, Hackensack, NJ 07601

UK office: 57 Shelton Street, Covent Garden, London WC2H 9HE

British Library Cataloguing-in-Publication Data
A catalogue record for this book is available from the British Library.

FROM ASSOCIATIONS TO RULES
Connectionist Models of Behavior and Cognition
Proceedings of the Tenth Neural Computation and Psychology Workshop

ISBN-13 978-981-279-731-5 (pbk)
ISBN-10 981-279-731-9 (pbk)

Printed in Singapore.

INTRODUCTION

The Tenth Neural Computation and Psychology Workshop (NCPW10) was held in Dijon, France, at the Laboratory for Learning and Child Development (LEAD) at the University of Burgundy. The organizers of the Workshop, Bob French and Xanthi Skoura-Papaxanthis, left its theme intentionally vague, even though it gravitated, as always, around connectionist models (preferably) of human behavior.

As every year, we attempted to combine a relatively small number of talks, with no parallel sessions or no posters, and lots of time for interaction between participants. Further, we took ample advantage of the fact that Burgundy is renowned for its *haute cuisine* and fine wine. This formula has proved to be a good one over the years, especially the explicit attempt to leave enough time free for researchers to talk to one another about their work. The Website for the conference can be found at: http://leadserv.u-bourgogne.fr/ncpw10. We are particularly indebted to Stéphane Argon for getting this site up and running for the Workshop, to Rosemary Cowell, who helped with the initial organization of paper presentations, and to John Bullinaria, one of the founders of the NCPW series of workshops, for his assistance time and again with tricky issues of all sorts.

The Workshop was supported by contributions from the CNRS, a European Commission grant (FP6-NEST 516542), the Conseil Régional de Bourgogne, and the University of Burgundy, as well as the participants' registration fees.

We have grouped the 18 papers making up this volume into essentially the same categories that we used during the conference – namely, High-level cognition, Language, Categorization and Visual Perception, Sensory and Attentional Processing.

Finally, we strongly hope that this tradition of a small, friendly meeting of neural net researchers interested in the modeling of cognition and behavior, begun 15 years ago in Bangor, Wales, will continue well into the future.

Robert French
Elizabeth Thomas
(editors)

CONTENTS

Section I

High-Level Cognition

A CONNECTIONIST APPROACH TO MODELLING THE FLEXIBLE CONTROL OF ROUTINE ACTIVITIES

NICOLAS RUH
Oxford Brookes University, Department of Psychology, Gipsy Lane, Oxford OX3 0BP

RICHARD P. COOPER
School of Psychology, Birkbeck, University of London Malet Street, London, WC1E 7HX

DENIS MARESCHAL
School of Psychology, Birkbeck, University of London Malet Street, London, WC1E 7HX

Previous models of the control of complex sequential routine activities are limited in that either a) they do not include a learning mechanism, or b) they do not include an interface with deliberative control systems. We present a recurrent network model that addresses these limitations of existing models. The current model incorporates explicit goal units and uses simulated reinforcement learning to acquire simultaneously both action sequence knowledge and knowledge of associated goals and hierarchical structuring. It is demonstrated that, in contrast to existing models, the revised model may both acquire task sequences and be controlled at multiple levels by biasing appropriate goal units.

1. Introduction

Routine activities are everyday behaviours that humans perform frequently without needing to pay attention to the task at hand, for example making the daily cup of breakfast coffee while still being half asleep or planning the day ahead. Norman & Shallice's [1] widely acknowledged dual systems theory of the control of action claims that two distinct systems contribute to the expression of complex sequential behaviour: One is an automatic conflict resolution system (Contention Scheduling, or CS), which selects from amongst the myriad of actions possible at any moment in time. CS is argued to function autonomously during the control of routine behaviour, i.e., when performing highly overlearned tasks such as one's daily breakfast routine. However, in deliberate behaviour (less familiar circumstances, novel tasks or dangerous situations), a higher level executive system (the Supervisory System, or SS) may exert control through modulation of CS. A useful analogy with regard to the relationship between these two systems is that of a horse (CS), which is doing all the actual work of locomotion and which may find the usual way home on its own, and a rider (SS), who can decide upon the path to follow in unusual circumstances.

From a computational point of view, this general picture poses two questions: (a) which kind of a system can mimic the horse's ability to perform such complex sequential tasks autonomously, and (b) how can the flexible interaction of horse and rider be captured. Previous computational accounts have focused on the first of these questions, claiming that either an Interactive Activation Network (IAN) approach [2] or a Simple Recurrent Network (SRN) architecture [3] is more suitable for modeling human routine behaviour and its breakdown. Both of these accounts, however, are models of fully routinised performance (CS) only and thus are unable to address the second question.

Our own recent empirical work [4] suggests that such an all-or-nothing approach to a specific task being either routine or not might be misleading. Our results instead support the conclusion that the dual systems interact in an extremely flexible manner, with the contribution of the SS, at every point in time, being dependent on factors such as the local complexity of the task and the amount of experience and structural overlap with this and similar tasks.

The amount and nature of the interaction between the two systems is necessarily complex as it may depend on the external situation (e.g., unexpected circumstances, perceived danger) and the internal state of the system (e.g., amount of experience, whether an error occurred earlier in the sequence). Furthermore, there might be variations in the strength and the level of control exerted by the SS at any given point in time. To illustrate why the deliberative system must be capable of different levels of control, consider the case of learning to drive. A novice typically attends to each single movement of every limb. With practice, one moves to a state where one may simply pay attention to where one is going when driving to an unfamiliar location – the expert driver needs only to attend to those aspects of driving that are not routine. Within the CS/SS theory, the SS would, in the latter case, merely influence higher level decisions such as which road to take at a junction, relying on a well trained CS to take care of the lower level sequences (breaking, indicating, etc.).

Error recovery, on the other hand, is more in line with the 'supervisory' nature of the SS helping out at critical points when triggered by unexpected input or recruited by some monitoring subsystem that indicates that things are not going as they should. The SS needs to 'know' how to modulate the CS in order to bring the system back into the routine execution of a given sequence.

Importantly, in all cases the SS is claimed to work by biasing CS at appropriate points in the sequence, rather than by taking over control completely. In functional terms this seems to require a massively redundant system that is able to flexibly switch on the higher level part (SS), on different levels (enforcing task goal, subgoals or single steps), either voluntarily or when triggered by some measure of conflict in the basic level system.

Building on the SRN approach [3], we will present an embedded connectionist model that employs explicit goal representations as an interface between SS and CS, thereby addressing the dynamic interaction between the horse and the rider in functional terms. The specific training regime ('simulated reinforcement learning') furthermore leads to a more flexible implementation of behavioural routines in terms of 'policies' rather than rigid individual sequences, thus allowing the model to pursue its goals even when faced with minor errors or irregularities in the environment.

2. Existing Models of Sequential Control

In search for a starting point for a model that shows progressive routinisation and allows control at multiple levels, we first consider the strengths and weaknesses of the two existing models of routine action. Both of these models implement the *de facto* prototypical routine task of preparing a hot beverage, either coffee (4 variants) or tea (2 variants, see Table 1). The model presented here will also be concerned with this very task. Note that we will use the term 'task level' (TL) when referring to full coffee or tea sequences, 'intermediate level' (IL) in reference to subtasks or subgoals such as the ones displayed in Table 1 and 'basic level' (BL) when speaking about the smallest meaningful chunks of actions that can be summarized under one goal (e.g., picking sth. up).

Table 1: The six valid variants of the beverage preparation task. Shown are intermediate level subtasks, each consisting of between four and ten individual action selection steps (e.g., add milk = fixate container – pick-up container – tear open – fixate cup – pour milk – put-down container – fixate spoon – pick-up spoon – fixate cup – stir). For more details please refer to [2] and [3].

coffee:	
c1:	add coffee grounds – add sugar from pack – add milk – drink
c2:	add coffee grounds – add milk – add sugar from pack – drink
c3:	add coffee grounds – add sugar from bowl – add milk – drink
c4:	add coffee grounds – add milk – add sugar from bowl – drink
tea:	
t1:	add teabag – add sugar from pack –drink
t2:	add teabag – add sugar from bowl – drink

Cooper & Shallice [2] describe an IAN model of routine sequential behaviour in which hand-coded, hierarchically organized action schemas correspond in a one-to-one fashion with nodes that have continuous-valued activations. Nodes may either excite subnodes according to the requirements of hierarchical organization, or inhibit nodes corresponding to competing schemas via lateral connections. In contrast, Botvinick & Plaut ([3], henceforth B&P) present an SRN model that acquires sequence information through repeated exposure to the six example routine sequences. The model learns a distributed representation of task context, which it uses to control sequential behaviour throughout a task.

In theory, nodes within the IAN model may be selectively biased by the SS, but as the IAN model provides no mechanistic account of learning, it can provide no mechanistic account of the transfer of control from SS to CS with practice. Conversely, while the learning of action sequences is straightforward in the SRN model, the fact that it employs a fully implicit representation of task context prevents a higher-level executive system from interfacing with the CS for the purpose of exerting explicit control, thus severely limiting the model's scope and extendibility. In addition, the way in which the SRN acquires its functionality is implausible. First, the set of exemplars on which the model is trained must be composed in such a way as to present an equally distributed choice of possible continuations at each subtask boundary [5]. If this assumption is not met, the lack of explicit control means that the branch with the lower frequency is inaccessible for the model, whose sole non-deterministic factor is the random initialization of the context layer. Second, the fact that the model is trained on exactly six valid sequences means that it is unable to deal with even the slightest deviations from these templates, for example if the initial state consists of anything but fixating on an empty cup and holding nothing, or if any of the objects encountered during the sequence is in an unexpected state (e.g., the sugarbowl is already open when first fixated).

3. The Goal Circuit Model

The SRN model provides an approach to learning hierarchical action sequences. This is essential when attempting to model the progressive routinisation of action sequences, but extending the model requires at least: a) a way to interface it with an executive system (SS) which can add executive control (bias) in the not fully routinised case; b) a more plausible training regime, taking into account the reusability of existing (sub)sequence knowledge and the progressive reduction of executive control with increasing practice; and c) the ability to reach the (sub)goal of a specific (sub)task in a more flexible manner, thus dealing with minor variations in the states of objects or the initial state of the system. The Goal Circuit (GC) model presented here attempts to extend B&P's SRN model along these lines.

3.1. *Goal Units and Simulated Reinforcement Learning*

It is implausible that we learn to make coffee by being guided repeatedly, in a step-by-step manner, through the entire task sequence by some external teacher (as in the SRN model). We suggest that the information we make use of is more like a high level description of subtasks: "boil water, add grounds, add milk and sugar, drink". There is a high level of agreement between people when asked what it takes to make a cup of coffee, and it is roughly these points that they mention [6].

The first insight of the GC model is that if a network has already learned to add grounds, sugar, milk, etc., and if it has control units corresponding to each of these goals, then it may be biased to perform more complex sequences (e.g., making a cup of coffee) by activating appropriate goal units in sequence. At the same time a higher-level goal unit could be trained to represent the transitions required for this task, thus making the detailed guidance at the lower level optional. The same argument applies to the lower level goals (i.e., acquisition of the add sugar routine), down to basic goals/actions (e.g., picking something up) that are invariant due to environmental constraints. In the GC model (see Figure 1; cf. Figure 3 of B&P) we thus added banks of goal units which encode goals at three levels of abstraction using a localist representation. The goal layers (input and predicted) included 11 basic level goals (BL: get one of 7 objects > open > add > stir > sip), 5 intermediate level goals (IL: add grounds, add teabag, add sugar, add milk, drink), and two task level goals (TL: make tea, make coffee).

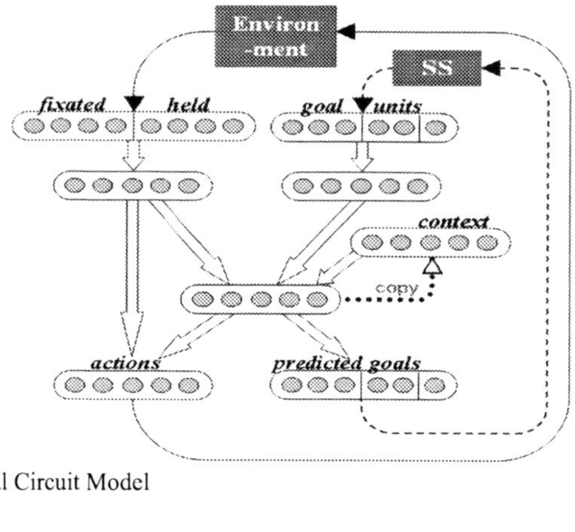

Figure 1: The Goal Circuit Model

The second insight of the GC model is based on previous work in which we demonstrated that reinforcement learning may be used to train a recurrent neural network to encode goal-directed action schemas [5]. Reinforcement learning involves allowing a system to generate random sequences of behaviour and providing positive reinforcement when those sequences achieve a desired goal. Systems of this sort learn to achieve their goals in a flexible way, so that goal achievement is independent of, for example, the initial state of the environment.

The problem is that most current implementations of reinforcement learning are limited to situations involving a single goal. Dissimilar policies (sets of optimal transitions) cannot be learned because the models lack the means to distinguish between rewards received for reaching different goals. To overcome this, the GC model was trained with an interactively generated set of sequences that aimed to include every possible way of, e.g., picking up an object, opening it, adding it to the beverage, etc. This included not only different starting states in terms of what is fixated/held initially, but also different states of the environment, e.g., sugarbowl is open or closed when fixated on. Each sequence was accompanied by its respective goal as an additional input, thus for all the different ways in which milk can be added the 'add milk' goal was supplied as an input from the goal bank. The full task sequences, however, still consisted of the six variants from Table 1, albeit with a random starting state. The rationale behind this 'simulated reinforcement' regime was to provide the network with a training set comparable to what a successful reinforcement learning model would have seen, i.e., all (or most) valid sequences that lead to a specific goal.

3.2. *Progressive Routinisation*

SRN models learn to use their context layer to "remember" where they are in a task sequence, but in the given task this memory is needed at very few points only – for example to prevent the model from stirring twice. The additional information provided by explicit goal units covers almost all cases where memory is needed, thereby making the context layer dispensable. Progressive routinisation, however, requires that the basic level system (CS) becomes increasingly independent from the activation of appropriate goal units provided by the higher level system (SS). In the present model this translates into requiring that the context layer gradually incorporates the functionality provided by the goal units. This process can be implemented by making the input from the SS (the goal units) unreliable during training, thereby encouraging the network's context layer to encode initial goals, to the point where a goal set at the first step only is able to provide the necessary bias for execution of the whole task sequence. In order to implement this unreliability, the input goal units being were turned off after the first step in 50% of all training sequences.

3.3. *Network Architecture and Parameters*

All other parameters were held as close as possible to B&P [3]. The perceptual input consisted of two 19 unit vectors, coding the objects currently fixated and held (e.g., cup, spoon, sugarpack, etc.). The output layer coded 19 possible actions in a localist manner (e.g., pick-up, open, pour, fixate spoon, etc.). All internal layers had 50 units. The additional recoding layers between input and hidden layer (cf. Fig. 1) are not strictly necessary, but help the network to perform more stably by balancing the varying levels of overall activation (i.e., different number of active units) in the input banks. A standard sigmoidal squashing function was employed for all units of the network. Training was terminated either by reaching a running average sum square error (with a window of 400 sequences) lower than 0.04 or after 100.000 sequences had been processed. Standard backpropagation was used with a learning rate of 0.02. Weights were initialized in the range of +/- 0.1. The target signal for the predicted goals consisted of all available goals, that is, one goal for each level when processing a TL sequence, the lower two level goals for an IL sequence and the basic level goal only when a BL sequence was being processed.

4. Results

The GC model is designed for flexibility, therefore testing the model needs to go beyond simply ensuring that it has learned the task. We tested the model in three different control modes, that is, in three different ways in which the SS might contribute to generation of a task sequence. The three control modes were:

Horse only mode: In this setting, the SS exerts no control on the basic level system beyond the first step. In psychological terms this setting reflects the situation where, after an initial intention to do something (e.g., making coffee) one directs attention elsewhere and does not pay attention to subsequent actions within the task. Obviously this can succeed only for fully routinized tasks.

Loop mode: In this setting, the self-generated goal predictions are directly copied into the goal units as inputs at the next time step. The effect is that the basic network gets biased towards using policies that correspond to the predicted goals, which will ultimately lead to these goal(s) being satisfied. This helps the model to stay on track in the presence of minor inconsistencies and as such implements a mechanism for automatic recovery from minor errors.

Homunculus mode: Appropriate goal units may be set at different points during sequence production by the SS. This setting reflects deliberate control of choices at different levels, depending on which goal units are activated. These deliberately activated goals are assumed to be the result of higher-level processes operating within the SS (e.g., problem solving, retrieval from explicit memory or following instructions).

During testing, all weights were frozen and the model's output was mediated via the environmental loop in order to generate the next input. The action with the highest activation was always executed.

4.1. *The Coffee or Tea Routine*

First, it is necessary to verify that the 'routine' performance of the GC model is comparable to B&P's SRN model. Table 2 shows the GC model's performance in horse only mode, thus not inputting any goals after the initial step.

Table 2: Each line corresponds to 100 trials with randomly initialized context layer. The left columns show the initial state, the middle columns show the number of task, intermediate and basic level sequences produced, an error is scored when no valid sequence was produced. The right columns shows a breakdown of the TL sequences into the different variants of coffee (c1, c2, c3, c4) and tea (t1, t2) making sequences. Note that the GC model is able to cope with different starting states.

fixated | held	initial goal	TL	IL	BL	err	c1	c2	c3	c4	t1	t2
cup | nothing	coffee	99	1	0	0	51	0	48	0	0	0
cup | nothing	tea	99	1	0	0	0	0	0	0	50	49
cup | nothing	no goal	0	3	92	5	0	0	0	0	0	0
cup | spoon	coffee	90	10	0	0	50	0	40	0	0	0
cup | spoon	tea	100	0	0	0	0	0	0	0	49	51
cup | spoon	no goal	0	6	94	0	0	0	0	0	0	0

Two differences are apparent in comparison to B&P model's performance. First, when no goal is initially set the GC model produces BL (or occasionally IL) sequences, rather than complete TL sequences. This is sensible: During training the model never sees cases with no initial goal. Its most frequent response of carrying out a basic level sequence reflects object affordances: when an object is frequently used in a certain way, perceiving it might be sufficient to trigger the associated behavior. Second, when the coffee goal is set the GC model develops order preferences. Here it prefers to add sugar before milk (sequences c1 and c3) rather than after (sequences c2 and c4). Again, the GC model's performance is sensible: in previous empirical research we found that most participants developed a preference for order when learning virtual versions of the coffee and tea tasks [4]. However, the GC model is able to produce less favoured sequences when appropriate IL goal units are activated during the task ('deliberate control', see below).

4.2. *Redundancy and Goal Directed Behaviour*

The testing described above can be replicated in the loop mode. Different variants are possible, such as exclusively feeding back goals from either level, or inputting all levels of predicted goals. All of these settings led to 100% correct production of the required task sequence. Critically, inputting self generated goals prevented the network from sometimes forgetting which goal it was aiming to achieve and thus stopping early (see Table 2). The fact that this worked at all levels of goals demonstrates the intended redundancy. Even a fully routinized task may be carried out with multiple degrees of control.

There is another situation in which the usefulness of the goal circuit can be demonstrated. One difficulty with the original SRN model is that it cannot cope with variations in the environment [7]. Our simulated reinforcement learning regime was aimed at teaching the GC model how to reach a goal (e.g., of adding sugar), rather than how to generate a specific sequence. Hence, the training set included instances of (IL) adding sugar when the sugar bowl was already open, requiring the model to omit pulling off the lid and to get the spoon for scooping instead. However, we deliberately excluded examples with the open sugar bowl in the TL context of making tea or coffee. Nevertheless, the model was able transfer its knowledge from intermediate level to task level sequences: when tested in an environment with an open sugar bowl, the model correctly omitted the opening routine in 25.5% of the tea tasks and 96% of the coffee tasks.

Figure 2 shows the activation of the model's output units in a successful tea trial. The sugar bowl is perceived in step 6 and, contrary to the model's expectation given its experience with this TL task, turns out to be already open. The expectation can be seen by the fact that the BL goal 'open' and, subsequently the action 'pull-off' are partially activated (although well below the usual activation level of 1.0). Active as well, though, is 'get/fixate spoon', because the model has previously experienced situations in which this was the appropriate reaction to perceiving the open sugarbowl. In the case shown, the correct action wins by a small margin, leading the model to overcome its difficulties and, with the help of the goal circuit, successfully complete the tea task. Note that the model finds its way back into predicting the correct task level goal after the temporary disruption (see Fig. 2, t1_t2 output in steps 8–11).

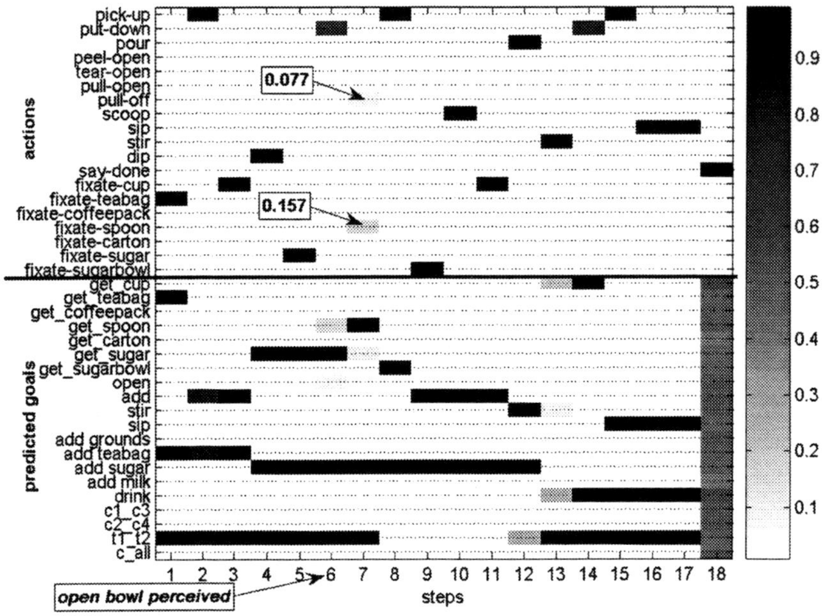

Figure 2: Successful trial of a tea sequence with an open sugarbowl.

The picture is very similar in the unsuccessful trials, except that the small trace of the random initialization of the context layer leads to the incorrect 'pull-off' action being the most active output in step seven. In principle, a small intervention by the SS, recruited by the model's uncertainty about which action to select in step seven, could tip the scale in favor of the correct continuation. Note as well that, if weights were not frozen, every occurrence of the open sugar bowl in the context of the tea task would lead the model to gradually include this situation in it's 'make tea' policy so that, with sufficient experience, action selection at step seven would no longer produce conflict.

The model's ability to deal with unexpected perceptual inputs demonstrates that the model has acquired flexible goal directed behavior. In contrast to the original SRN of B&P it is capable of taking minor variations of its routine in stride by applying knowledge acquired in the context of smaller subtasks.

4.3. *Deliberate Control*

The goal circuit implements one of the less demanding functionalities of the SS. It basically serves as a simple monitoring and clean-up mechanism in well-learned tasks. This represents by no means the limit of the possible control that the SS should be able to exert. The SS might come to the decision to support a specific goal in many other ways (e.g., instruction, imitation, problem solving). In homunculus mode, these more sophisticated functionalities of the SS are substituted by the experimenter actively setting goal units during task performance. The aim of this test mode was to establish that the interface with the SS was functional, even though the SS itself was not fully implemented.

First, we established that it was indeed possible to elicit a full task sequence by activating intermediate level goals in the appropriate order. With such tight control the model could be made to produce all valid task sequences, dependent on the order of the activated goals. It was also possible to guide the model trough as yet unknown variants, such as adding sugar twice. In most cases, the model was able to finish off the routine on its own, e.g., it was not necessary (though not harmful either) to activate the 'add milk' and 'drink' goals after the model had been forced to add a second portion of sugar in the coffee routine.

Using explicit goal units, it is further possible to ask the GC model to produce any basic level or intermediate level sequence from all possible starting states and random values of the context layer units. This seemingly small observation is important because it indicates that in the event of an error, independently of what exactly has gone wrong (random history (context layer) and random perceptual state), the model will be able to recover and find its way back into satisfying the goal currently input – provided the SS is able to determine what this goal should be. In some cases a mechanism as simple as the goal circuit may be able to provide this current goal or subgoal (as in the last section). In others, it may be necessary to recruit more sophisticated functionalities of the SS (e.g., problem solving, comparison to explicit procedural knowledge, etc., see [8]), but as the problem is always indicated by uncertainty and conflict in the goal prediction layer, this necessity is easily detected. The more sophisticated functions of the SS can then activate the appropriate (sub)goal, thereby guiding the basic system back into the intended path and, usually, enable further execution in routine mode.

5. Discussion

The GC model presented offers a computationally explicit account of the interplay of the two systems held to pertain in the execution of complex sequential (routine) activities. The combination of a more plausible training paradigm and explicit goal representations in a connectionist framework results in a model that exhibits extremely flexible behaviour, addressing not only the progressive routinisation of frequently performed action sequences but also how the higher-level control system (SS) can be used to deliberately guide behaviour when necessary. The goal circuit and the context layer provide alternative ways of incorporating context information into the mapping from perceptual input to action output. The level of the context layer's contribution depends on the network's experience with similar sequences; the model learns to maintain past information and to use it to guarantee the integrity of sequences when necessary. The information employed may include past actions (e.g., using the memory of just having stirred to avoid doing it again) and/or past goals (e.g., maintaining the 'make tea' goal to suppress actions involved in making another beverage). The explicit goal units, conversely, are used to enforce the correct expression of sequences before the context layer has learned to do so (in novel tasks) or during error recovery (when the model is off track and must be guided back into a sequence).

The GC model reaches a level of (routine) performance that is comparable to the B&P model. Importantly it does this while employing a training algorithm (standard backpropagation) that is local in time. This is not only an advantage in terms of biological plausibility; it furthermore opens up the possibility to train the model as a settling network. The advantage of having the network settle within individual steps would be the opportunity to directly compare settling times to the action selection latencies of participants performing this task [4].

Another extension of interest concerns the use of reinforcement learning. The current model attempts to simulate reinforcement learning by employing a training corpus that includes all the valid sequences a successful reinforcement learning model is likely to have discovered. The training itself though is done in a fully supervised manner. Standard reinforcement learning models do not address the problem of multiple independent reinforcement signals (goals), or of switching between policies to pursue different goals. In a full reinforcement learning implementation, the model needs not only discover viable policies to satisfy multiple goals by exploration/exploitation, but at the same time it must learn to distinguish between policies that lead to the different goals, thus providing the SS with the means to address each goal independently. Recent modeling work [9] has made inroads into this problem of learning to control – or, in the context of our analogy, of teaching the rider how to ride. While these models thus far are concerned with less complex tasks, it might be possible to transfer some of the computational solutions to the GC model, thus potentially alleviating the need to enforce an appropriate goal hierarchy via the training set.

What we have attempted to capture with this model is how a hypothesised SS could interface with a distributed CS, and not the detailed workings of the SS itself. In testing the model we make use of some simple but efficient ways to connect the predicted goals to the goal input (the goal circuit). We hold no strong theoretical commitment to this precise looping architecture. An attractive alternative could be deduced from a recent model of working memory [9]. In this model, independent parallel loops connect an Adaptive Critic, held to be localised in the basal ganglia, to prefrontal areas that are widely recognized to serve executive functions. These parallel loops provide a gating signal that indicates when it is useful to switch to a new context that is subsequently maintained. While the maintained/gated information in this model is a perceived stimulus, a similar approach could also be applied to the gating of goals.

References

1. DA Norman and T Shallice in Consciousness and Self Regulation, eds. R Davidson, G Schwarz and D Shapiro (Plenum, New York, 1986).
2. R. P. Cooper and T. Shallice, *Cogn. Neuropsych.* **17**, 297 (2000).
3. M. M. Botvinick and D. C. Plaut, *Psych. Rev.* **111**, 395 (2004).
4. N. Ruh, R. P. Cooper and D. Mareschal, *Proc. CogSci'05*, 1889 (2005).
5. N. Ruh, R. P. Cooper and D. Mareschal, *Proc. AKRR'05*, (2005).
6. G. W. Humphreys and E. M. Forde, *Cogn. Neuropsych.* **15**, 771 (2000).
7. R. P. Cooper and T. Shallice, *Psych. Rev.* **113**, 887 (2006).
8. T Shallice in Attention and Performance XXI, eds. Y Munakata and MH Johnson (University Press, Oxford, 2006).
9. R. C. O'Reilly and M. J. Frank, *Neural Computation*, **18**, 283 (2006).

ASSOCIATIVE AND CONNECTIONIST ACCOUNTS OF BIASED CONTINGENCY DETECTION IN HUMANS*

SERBAN C. MUSCA, MIGUEL A. VADILLO, FERNANDO BLANCO AND HELENA MATUTE

Laboratorio de Psicología del Aprendizaje, Universidad de Deusto, 24, Avenida de las Universidades, 48007 Bilbao, Spain

Associative models, such as the Rescorla-Wagner model (Rescorla & Wagner, 1972), correctly predict how some experimental manipulations give rise to illusory correlations. However, they predict that outcome-density effects (and illusory correlations, in general) are a preasymptotic bias that vanishes as learning proceeds, and only predict positive illusory correlations. Behavioural data showing illusory correlations that persist after extensive training and showing persistent negative illusory correlations exist but have been considered as anomalies. We investigated what the simplest connectionist architecture should comprise in order to encompass these results. Though the phenomenon involves the acquisition of hetero-associative relationships, a simple hetero-associator did not suffice. An auto-hetero-associator was needed in order to simulate the behavioural data. This indicates that the structure of the inputs contributes to the outcome-density effect.

1. Introduction

Perceiving contingency between potential causes and outcomes is of crucial importance in order to understand, predict, anticipate and control our environment. However, there is little agreement on the mechanisms that underlie this ability, and the research on human contingency perception is in its flourishing years.

As with many other cognitive phenomena, one way to gain a better understanding of the ability under scrutiny is to find variables that affect it in a systematic way, be able to predict their influence, and finally comprehend why these variables do have an effect. Personality or mood variables (e.g. Alloy & Abramson, 1979), the valence of the outcome (e.g. Alloy & Abramson, 1979; Aeschleman, Rosen & Williams, 2003), the density of the cue/response (e.g.

* Support for this research was provided by Grant SEJ2007-63691/PSIC from the Spanish Government and Grant SEJ406 from Junta de Andalucía. Fernando Blanco was supported by a F.P.I. fellowship from Gobierno Vasco (Ref.: BFI04.484).

Allan & Jenkins, 1983; Matute, 1996), the density of the outcome (e.g. Alloy & Abramson, 1979; Allan & Jenkins, 1983; Matute, 1995) are all factors that have an influence on the ability to correctly perceive contingency in humans.

In the following, after describing the general methodology used in behavioural experiments that study contingency perception, we will focus on the influence that density of the outcome has on a judgment of contingency. We will present the widely accepted associative account of Rescorla and Wagner (1972) and also behavioural data that are not accounted for by this model. We will then present simulations conducted with two different neural network models designed to encompass those behavioural results not accounted for by the associative model, and the surprising results the simulations yielded.

1.1. *Studies of Contingency Judgment*

Experimental studies of contingency judgment in humans generally involve the use of a 2-phase task. During the first phase (the training), covariational information is given to the participants in successive trials. In each trial a cue (e.g. ingestion of strawberries) is either present or absent, and the outcome of interest (e.g. allergic reaction) either occurs or does not occur (cf. Table 1). Trials where both the cue and the outcome are present (*a* trials), trials where the cue is present and the outcome absent (*b* trials), trials where the cue is absent and the outcome present (*c* trials), and trials where both the cue and the outcome are absent (*d* trials) are presented in random order for a total of (a+b+c+d) trials.

Table 1. Trial types that make up the covariational information that is given to the participants in contingency judgment experiments.

		Outcome (allergic reaction)	
		present	absent
Cue (eaten strawberries)	present	a	b
	absent	c	d

In the subsequent test phase participants are to judge the degree of the causal relationship between the cue and the outcome (e.g. to what degree they think the ingestion of strawberries is the cause of the allergic reaction).

Of course, this is just a general outline, and many variants of the task exist. For instance, the subjective contingency can be assessed throughout the learning phase by presenting the participants with the cue and asking them to predict what the outcome would be before displaying the actual outcome. In another task that has been used extensively, during the training phase the cue (present/absent) is replaced by the participant's response (i.e. response/no response).

Participants generally get things right, but under certain conditions participants' judgments diverge from the ideal judgment one is expected to give based on the objective covariational information presented during the training phase (López, Cobos, Caño, & Shanks, 1998). However, in order to observe this discrepancy, one has to dispose of a measure of the ideal judgment expected.

An objective measure has to take into account both the probability of a present outcome when the cue was present — that is p(O|C) — and the probability of a present outcome when the cue was absent — that is p(O|noC). Indeed, the fact that the outcome is present when the cue is present does not mean that the cue is the *cause* of the outcome if the outcome is present just as many times when the cue is not present. Based on this reasoning, the ΔP index was proposed by Jenkins & Ward (1965; see also Allan, 1980; Cheng & Novick, 1992) as a measure of contingency:

$$\Delta P = p(O|C) - p(O|noC) = a/(a+b) - c/(c+d) \qquad (1)$$

This index has a value of 0 if the presence of the cue is not the cause of the presence of the outcome, that is, if the presence of the outcome is not contingent on the presence of the cue.

1.2. *Illusory Correlation*

As the ΔP formula hints, to one ΔP value may correspond many different distributions of trial types (see Table 1 for the four trial types), with different cue probability — *cue density*, p(C) = (a+b)/(a+b+c+d) — and/or outcome probability — *outcome density*, p(O) = (a+c)/(a+b+c+d).

As noted in the introduction, it is documented that these densities (among other variables) do bias the perceived contingency in the sense that they incorrectly affect participants' judgment of contingency. *Illusory correlation* refers to the phenomenon whereby in a noncontingent situation (i.e. a situation of stochastic independence between cue and outcome) participants incorrectly perceive a contingency between the cue and the outcome. The contingency between cue and outcome perceived by the participants is illusory because considering the covariational information supplied to the participants ΔP is nil.

In the following we will consider to a greater extent one of the possible causes of illusory correlation, the outcome density.

2. Outcome-density Effect

The outcome-density effect is an illusory correlation that has its roots in the probability of occurrence of the outcome (or *outcome density*). Though participants' judgments of contingency should not differ while the contingency is kept constant, it is documented that participants' judgments of contingency differ following the probability of the outcome, p(O). This bias was called outcome-density effect (e.g. Alloy & Abramson, 1979; Allan & Jenkins, 1983; Matute, 1995). For instance, while ΔP is nil both for $a = 15$, $b = 5$, $c = 60$, $d = 20$ and for $a = 5$, $b = 15$, $c = 20$, $d = 60$, p(O) is of 0.75 in the former case, and of 0.25 in the latter. In the example given here, were the participants to rate the contingency as higher when p(O) = 0.75 than when p(O) = 0.25 one would speak of an outcome-density effect.

2.1. *An Associative Account: The Rescorla-Wagner Model*

The Rescorla-Wagner model (hereafter RW) proposed by Rescorla & Wagner (1972) is one of the most widely used associative models when it comes to simulate how people learn to associate potential causes and effects (here, cue and outcome). Sutton & Barto (1981) have shown that it is formally equivalent to the delta rule (Widrow & Hoff, 1960) used to train two-layer distributed neural networks through a gradient descent learning procedure. In the RW model, the change (ΔV_C^n) in the strength of the association between a potential cue C and a potential outcome after each learning trial takes place according to the equation:

$$\Delta V_C^n = k \cdot (\lambda - \Delta V_C^{n-1}), \tag{2}$$

where k is a learning rate parameter that reflects the associability of the cue, α, and that of the outcome, β ($k = \alpha \cdot \beta$ in the original RW model); λ reflects the asymptote of the curve (which is assumed to be 1 in trials in which the outcome is present and 0 otherwise), and ΔV_C^{n-1} is the strength with which the outcome can be predicted by the sum of the strengths that all the cues that are present in the current trial had in trial *n-1*.

The RW model correctly predicts that outcome density manipulations give rise to illusory correlations. This is illustrated in Figure 1, by manipulating the outcome density in a case where the total covariational information corresponds to a noncontingent situation (i.e. $\Delta P = 0$). The parameter k was set to 0.3 for the cue and to 0.1 for the context. We run 3000 replications. As can be seen in Figure 1, the associative strength developed between the cue and the outcome, which corresponds to an illusory correlation in this case (because ΔP is nil), is stronger and more long-lasting for the case when the outcome density is higher.

The RW model correctly predicts and simulates a large set of associative learning phenomena (for a review see Miller, Barnet, & Grahame, 1995; López et al., 1998). However, in the following we will focus on a set of data that are not accounted by this model, and see why the characteristics of the model make it impossible for it to simulate these behavioural data.

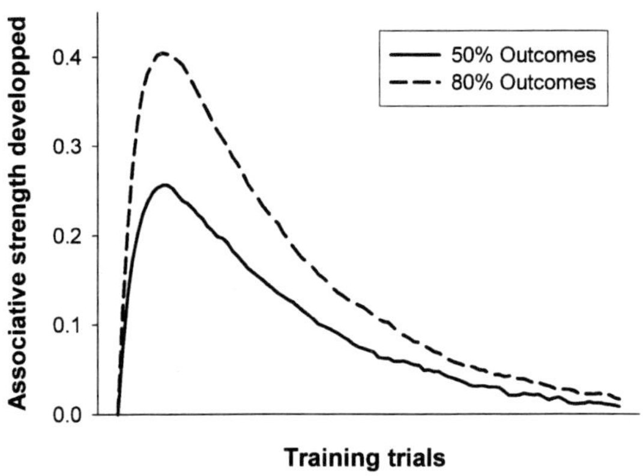

Figure 1. Illusory correlation (associative strength developed between the cue and the outcome) in the RW model in a noncontingent situation (see text for details).

The simulation presented above hints two characteristics of the RW model. One of them is that it predicts that outcome-density effects are a preasymptotic bias that should vanish as learning proceeds (e.g. López et al., 1998; Shanks, 1995). Indeed, from inspection of Figure 1 one can see that even when the outcome density is high, with enough training the associative strength developed between the cue and the outcome finally goes down to zero.

Another characteristic of the RW model is that the associative strength developed between the cue and the outcome is positive, so that with a low outcome density (e.g. 20%) the model will not yield a negative illusory correlation but a positive one.

While data contradicting these predictions of the model were at first scarce and were considered as anomalous, behavioural results at odd with these two RW model characteristics accumulated over the years. For instance, there is evidence that the outcome-density effect sometimes persists even with extensive training (Shanks, 1987) and that it does not disappear but, on the contrary, become stronger with more training trials (Allan, Siegel & Tangen, 2005). Also, some experiments with noncontingent situations that comprise a condition of low outcome density yielded a negative illusory correlation (Shanks, 1985, 1987; Allan et al., 2005; Crump, Hannah, Allan, & Hord, 2007).

In view of these results that the RW model cannot simulate and account for, the two aforementioned characteristics of this model appear as limitations. In the following we consider another type of modelling, one using simple distributed artificial neural networks. Using this more powerful metaphor, we investigated what the simplest connectionist architecture should comprise in order to encompass these results.

2.2. *Connectionist Simulations: What is the Minimal Model?*

The delta rule (Widrow & Hoff, 1960) on top of being equivalent to the learning rule used by the RW model is also the ancestor of the generalized delta rule (Rumelhart, Hinton & Williams, 1986) that once discovered in the late 1980's gave rise to the connectionist revolution we all know of. The generalized delta rule allows training networks of more than two layers, some of which have a complex nonlinear behaviour. These are more powerful simulation tools than the RW model, but when using this class of models one has to keep the simulation model to a minimum and analyse the constraints and the degrees of freedom it allows for.

In accordance with this principle, and because of our previous work and inclinations (Musca, 2005; Musca & Vallabha, 2005) we chose to tackle the problem at hand with 3-layer distributed neural networks trained with a backpropagation learning algorithm that minimizes the cross-entropy cost function (Hinton, 1989). Because the problem involves the learning of cue-outcome pairs and the structure of the outputs is of interest (the outcome density is a property of the outputs), the minimal model to be used is a hetero-associator (Bishop, 1995). For both the hetero-associator initially considered and the augmented auto-hetero-associator used afterwards (see below), the learning rate was set to 0.1, the momentum to 0.7 and the activation of the bias cell to 1. For both types of architecture used, 50 replications were run, with matched connection weights between the two conditions that were contrasted (i.e. low vs. high outcome density).

2.2.1. *Translating the Problem into "Neural Networks Language"*

An important part of the modelling is the translation of the problem into neural networks language. This involves creating a training set, that is, choosing the input and output vectors and the way they are related one to another.

One element of importance is that because of its mode of functioning a neural network cannot learn at the same time (i.e. as part of the same problem) something and its contrary. In other words, two trials such as "cue present-outcome present" and "cue present-outcome absent" cannot coexist in a training base, unless they occur in a different context. This is a supposition that has to be done, but we think it is a sensible one. After all, you will burn your fingers when touching an oven or not depending on the context, that is whether it has been used and is hot or whether it has not been used for a long time, but you will not be able to tell whether your fingers will be burned or not when touching the oven if you do not dispose of the context information. When a hetero-associator neural network is trained, its task can be understood just as this: give the right output given the input at hand. So, if the input at hand gives different outputs depending on the context where it occurs, this context must be specified. With these considerations in mind, we decided that the training base will have as many different contexts as training trials (i.e. that no training exemplar had the same context as another training exemplar). While this is not the most economical solution it has the advantage of avoiding possible biases due to the sharing of context between trials.

The training base comprised 100 training exemplars. The input of each exemplar is a 102-component vector made of two parts. The first part is a 100-component context vector that contains only one *1* component and 99 *0* components in such a way that the context vectors of all the training exemplars are orthogonal. The second part of the input vector is a 2-component vector that codes for the cue, with *1 0* being cue present and *0 1* being cue absent. The output vectors are 2-component vectors that code for the outcome, with *1 0* being outcome present and *0 1* being outcome absent.

The dependent variable that we used, which we call "contingency estimation" is computed according to the following reasoning. After training, the network is probed without any context and the activation of the first output node is recorded, first with the *cue present* input (i.e. *1 0*) and then with the *cue absent* input (i.e. *0 1*). The network's *contingency estimation* is the difference in activation between the first recording and the second recording. Inspired by the ΔP index (see Equation 1), this index is computed as:

$$\text{Contingency estimation} = \text{activation}\ (O|C) - \text{activation}\ (O|noC) \qquad (3)$$

The covariational information given to the networks corresponds to a noncontingent situation, with an outcome density that was varied with two values, 40% (low) and 60% (high)[a]. In terms of types of trials (see Table 1) the low outcome density condition corresponds to $a = 32$, $b = 48$, $c = 8$, $d = 12$, and the high outcome density condition corresponds to $a = 48$, $b = 32$, $c = 12$, $d = 8$.

2.2.2. *Three-layer Hetero-associative Network*

Starting with random connection weights — uniformly sampled between -0.5 and 0.5 — the 3-layer network with 102 input units, 10 hidden units and 2 output units was trained with the abovementioned parameters.

The dependent variable contingency estimation was computed at three points during training: when the root mean squared error on the training set (RMS Error) was of 0.1, of 0.01 and of 0.001, which corresponds roughly to 10, 20 and respectively 100 training epochs. As the results depicted in Figure 2 show, be it with little or extensive training the hetero-associative network failed to exhibit the expected outcome-density effect.

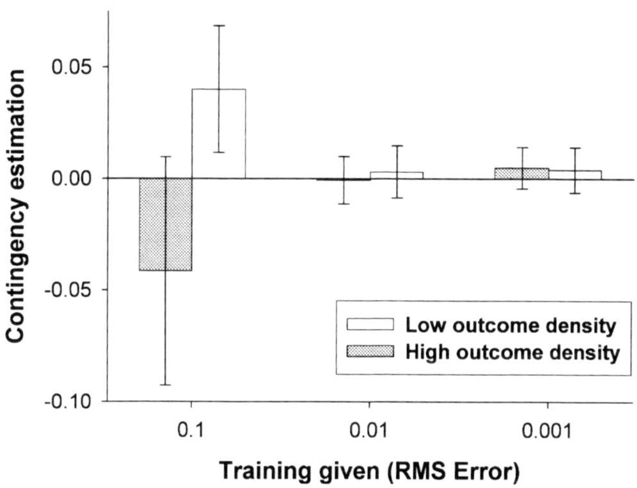

Figure 2. Results of the simulation with a hetero-associative network: the outcome-density effect is not simulated (whiskers represent .95 confidence interval limits).

The failure of the hetero-associator to produce the outcome-density effect is very surprising, because this kind of network builds its internal representations by taking into account the structure of the outputs, which is exactly what is manipulated when we use two different outcome densities. However this

---

[a] We chose outcome density values close to 50% so as to avoid a ceiling effect in the simulations.

surprising pattern of results seems to be robust, so we had to understand why it occurs. Could it be that what is called outcome-density effect is in fact not an effect that comes only from the density of the outcome? Moreover, could it be that the density of the cue plays a role in what is called outcome-density effect?

2.2.3. *Three-layer Auto-hetero-associative Network*

In order to answer these questions we resorted to a slightly augmented architecture, an architecture that takes into account when building its internal representations both the structure of the outputs (in this it is a hetero-associator) and that of the inputs (in this it is an auto-associator). Thus it is an auto-hetero-associator that we used in this second neural network simulation.

An auto-hetero-associator is a 3-layer network very similar to a hetero-associator, but its output layer contains not only the hetero-associative units but also other units, corresponding to the input units. Its task is both to associate the current input with the current hetero-associative target and to re-create at the output layer the current input.

Starting with random connection weights — uniformly sampled between -0.5 and 0.5 — the 3-layer auto-hetero-associator with 102 input units, 10 hidden units and 104 output units (the same 2 hetero-associative units as in the previous simulation, plus 102 auto-associative units used to reproduce the 102 input units) was trained with the same parameters as the hetero-associator in the previous simulation.

The dependent variable contingency estimation was computed at five points during training, three corresponding to those used in the previous simulation, and 2 complementary intermediate ones: when the RMS Error was of 0.1, of 0.075, of 0.01, of 0.005 and of 0.001, which corresponds roughly to 10, 40, 200, 350 and respectively 1500 training epochs.

Inspection of Figure 3 allows noticing two remarkable findings. First of all, both positive and negative outcome-density effect are obtained. This is compatible with the behavioural results that exist in the literature where both positive and negative illusory correlations have been found when manipulating the outcome density.

Secondly, the effects do not appear immediately but only after quite a considerable amount of training. However they do not vanish but become stronger when more training is given. And, though this may be artefactual and should be investigated at more length, a negative outcome-density effect seems to be obtained easier, (i.e. before, with less training) than a positive one, a result that has been found in humans by Crump et al. (2007). The pattern of results found in this simulation is compatible with the results of Shanks (1987) where a negative outcome-density effect was still present after very extensive training and with results of Allan et al. (2005) and Shanks (1985) showing that illusory correlation increases with training (but see López et al., 1998 for divergent results).

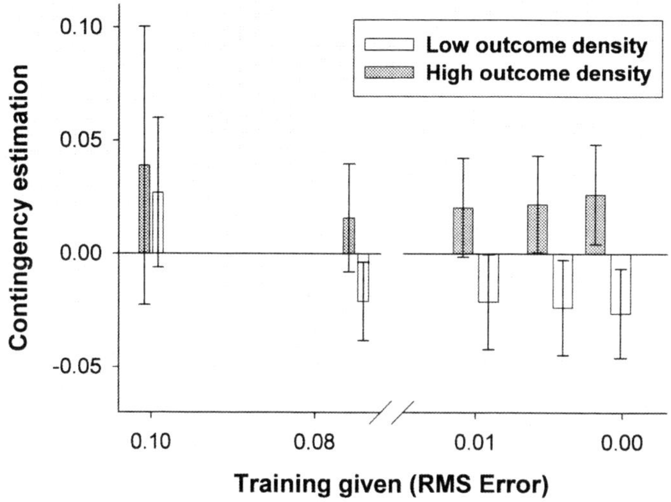

Figure 3. Results of the simulation with an auto-hetero-associative network: the outcome-density effect is simulated (whiskers represent .95 confidence interval limits). See text for details.

3. Conclusion

When trying to simulate those outcome-density effects found in humans the Rescorla-Wagner model cannot account for we had recourse to a distributed artificial neural network that is known to be sensitive to the structure of the outputs, that is, to the density of the outcome.

The simulation with this kind of neural network, a hetero-associator, failed to produce the expected outcome-density effects. This failure, in artificial neural networks terms clearly has only one implication: the so-called outcome-density effect is not an effect of the density of the outcome *per se*. Were it the case, the outcome-density effect would have been simulated with this class of networks.

In a second simulation we used an auto-hetero-associator, a type of network that takes into account both the structure of the outputs and of the inputs. With this distributed artificial neural network we were able to simulate the negative outcome-density effects that exist in the literature. Thus this model encompasses some important results that cannot be explained by the RW model. However, one must keep in mind that the RW model can explain data in the literature that an auto-hetero-associator could not simulate without complementary suppositions.

Moreover, quite some training was needed to the auto-hetero-associator before the effects appeared. Thus one possible explanation of the fact that such results are scarce with humans is that the training phase of most behavioural experiments is never extensive. Once the effects had appeared, the more training was given the stronger the effects became. Taken together, these results may point at a possible explanation of why some authors have found with a limited number of trials that the positive outcome-density effect dropped down to zero (see the apparent decrease in the outcome-density effect in Figure 3 when the RMS Error goes from 0.1 to .075).

In conclusion, it seems that the minimal distributed artificial neural network that is needed to account for the illusory correlations that are generated by the manipulation of the density of the outcome is an auto-hetero-associator. While this model seems powerful enough to account for a lot of extant data in the literature, we think its main interest lies in the predictions it makes. For instance, whether the so-called outcome-density is related not only to the density of the outcomes but also to the structure of the inputs, this could be checked in simulation work that manipulates both cue-density and outcome-density.

References

1. S. R. Aeschleman, C. C. Rosen and M. R. Williams, *Beh. Proc.* **61**, 37 (2003).
2. L. G. Allan, *Bul. Psychon. Soc.* **15**, 147 (1980).
3. L. G. Allan and H. M. Jenkins, *Learn. and Motiv.* **14**, 381 (1983).
4. L. G. Allan, S. Siegel and J. M. Tangen, *Learn. & Behav.* **33**, 250 (2005).
5. L. B. Alloy and L. Y. Abramson, *J. of Exp. Psych.: Gen.* **108**, 441 (1979).
6. C. M. Bishop, (1995). *Neural Networks for Pattern Recognition.* Oxford: Oxford University Press.
7. P. W. Cheng and L. R. Novick, *Psych. Rev.* **99**, 365 (1992).
8. M. J. C. Crump, S. D. Hannah, L. G. Allan and L. K. Hord, *Quart. J. of Exp. Psych.* **60**, 753 (2007).
9. G. E. Hinton, *Artif. Intell.* **40**, 185 (1989).
10. H. M. Jenkins and W. C. Ward, *Psych. Monograph.* **79**, 1 (1965).
11. F. J. López, P. L. Cobos, A. Caño and D. R. Shanks, *In* M. Oaksford & N. Chater (Eds.) Oxford: Oxford University Press, 314 (1998).
12. H. Matute, *Quart. J. of Exp. Psych.* **48B**, 142 (1995).
13. H. Matute, *Psych. Sci.* **7**, 289 (1996).
14. R. R. Miller, R. C. Barnet and N. J. Grahame, *Psych. Bul.*, **117**, 363 (1995).

15. S. C. Musca, *In* A. Cangelosi, G. Bugmann & R. Borisyuk (Eds.), Singapore: World Scientific, 367 (2005).
16. S. C. Musca and G. Vallabha, *In* B. G. Bara, L. Barsalou, & M. Bucciarelli (Eds.) Mahwah, NJ: Lawrence Erlbaum Associates, 1582 (2005).
17. R. A. Rescorla and A. R. Wagner, *In* A. H. Black & W. F. Prokasy (Eds.), New York: Appelton-Century-Crofts, 64 (1972).
18. D. E. Rumelhart, G. E. Hinton and R. J. Williams, *In* D. E. Rumelhart and J. L. McClelland (Eds.), Cambridge, MA: MIT Press, 318 (1986).
19. D. R. Shanks, *Mem. & Cogn.* **13**, 158 (1985).
20. D. R. Shanks, *Learn. and Motiv.* **18**, 147 (1987).
21. D. R. Shanks, *Quart. J. of Exp. Psych* **48A**, 257 (1995).
22. R. S. Sutton and A. G. Barto, *Psych. Rev.* **88**, 135 (1981).
23. G. Widrow and M. E. Hoff, *In* Convention Record of the Western Electronic Show and Convention, New York: IRE, 96 (1960).

ON THE ORIGIN OF FALSE MEMORIES: AT ENCODING OR AT RETRIEVAL? – A CONTEXTUAL RETRIEVAL ANALYSIS

EDDY J. DAVELAAR

School of Psychology, Birkbeck, University of London, Malet Street WC1E 7HX, London, United Kingdom

In the Deese/Roediger-McDermott false memory paradigm, participants produce many false memories. However, debates center on the question whether false memories are due to encoding or retrieval processes. Here, I present a novel way of analyzing data from a free recall paradigm, based on recent theoretical developments in the memory literature and apply this analysis to two experiments. The results suggest that the earliest process leading to false memories operate at encoding, but that the false memory itself does not affect further encoding processes.

1. Introduction

The healthy human memory system stores and retrieves a myriad of experiences of our daily lives, but is not perfect; not all retrieved memories are representations of truly experienced events. In this paper, I will focus on the memory illusion that is commonly referred to as a "false memory". One difficulty in investigating the psychology of false memories, that is their low rate of occurrence, has been overcome by the introduction of the Deese/Roediger-McDermott (DRM) false memory paradigm (Roediger & McDermott, 1995). In this paradigm, a participant is presented with a list of words, such as *bed*, *rest*, *awake*, and *tired* that are all related to a not-presented "lure" word (*sleep*). When the participant's memory for the list-words is tested, the non-presented word is falsely recognized or falsely recalled with high probability and high subjective confidence. The high occurrence of false memories makes this paradigm useful for investigating the origins of false memories in the laboratory.

In this paper, with "the origins of false memories" is meant the earliest cognitive operation that leads to a false memory. This operation need not be conscious or deliberate, but sufficient to produce a false memory if no other cognitive process intervenes. In particular, I will try to make the argument that with recent developments in the memory literature, we are in a position to address the question whether the earliest cognitive operation happens during the encoding of the list items or at retrieval of those items. I will first review some illustrative studies that highlight some of the difficulties that face researchers. After this, I will review new developments in the memory literature related to contextual retrieval and suggest that a novel type of data analysis can be extracted from these developments. I will apply this analysis on results of two experiments to address the main question and show the utility of this approach. Under the surface of this paper lies the opinion that detailed computational models of memory phenomena allow us to develop new theoretically-motivated analytical techniques that a non-modeler could employ in their own research without the need to master the computational modeling skills.

2. The Deese/Roediger-McDermott paradigm

As described above, the DRM paradigm consists of presenting a list of words that are all related to a non-presented word and when testing the memory for the studied words, looking at the occurrence of the critical lure word in the participant's report. As this paradigm is easy to employ in a psychological laboratory, it may come as no surprise that many variants of this task have been used, manipulating a range of variables, and studying different populations (e.g., neuropsychological patients, children, and elderly). For a recent review, the reader is referred to the book by Gallo (2006). In this section, I will briefly mention a few theoretical accounts of false memory and present the necessary terminology needed for the later sections. Then, I will review work that relates to the central question of this paper; do false memories originate during encoding or during retrieval of the words?

2.1. *Theoretical accounts of false memories*

Gallo (2006) summarized a number of theories that have been proposed to explain performance in the DRM paradigm. He noted that the theoretical accounts tend to focus on a small subset of the known data, particularly from the variant using recognition as the memory test (which leads to the bias in many theories to use a matching metaphor). In recent years, a number of computational theories have been developed that specifically address some of the added complexities in memory recall (Davelaar, et al., 2005; Howard & Kahana, 2002), allowing a sizeable body of DRM work (using a recall test) to be incorporated in theory development (see e.g., Kimball, Smith & Kahana, 2007).

Gallo (2006) distinguishes between decision- and memory-based accounts. Decision-based theories propose that a memory signal for the lure is absent and that the high false alarm rate is due to factors operating at retrieval such as criterion shifts in recognition (Miller & Wolford, 1999). Memory-based accounts, however, assume that the critical lure elicits a memory signal (which is stronger than unrelated lures). This signal could be due to activation of an actual memory trace for the lure or due to activation of studied items that are related to the critical lure. These are retrieval- and encoding-accounts respectively.

Underwood (1965) suggested that encoding of the various associates of the critical lure causes the representation for the lure to become activated. He called this indirect activation, an implicit associative response (IAR) and it has a major influence on recent theorizing (Roediger, Balota & Watson, 2001). In this account, the lure is activated and the participant becomes consciously aware of the lure and its associative nature. A strong alternative is the fuzzy-trace theory (Brainerd & Reyna, 2002), which proposes that during encoding a verbatim trace and a gist trace are created. At retrieval, the relative contribution of the two traces governs the amount of false memories, as the critical lure only overlaps with the gist trace. These two theories are by no means the only ones, but they do form the background against which the encoding/retrieval aspect can be addressed.

Within the IAR-account, the critical lure is activated during the encoding of related items (and this episodic trace leads to a false recall during retrieval) or is activated during the retrieval of related items (a form of retrieval-induced priming). Within the fuzzy-trace theory, the critical lure is only activated during retrieval and only to the extent that the gist trace is used in recall (no retrieval-induced priming). This means that the origin of false memories can be placed during encoding if it can be demonstrated that lure-specific information has been activated and encoded in an episodic trace. A retrieval account could predict that the critical lure is reported after a few strong associates if the false memory is due to retrieval-induced priming.

2.2. *DRM findings related to encoding/retrieval*

Seamon and colleagues (Seamon, et al., 2002) investigated the IAR-account using an overt-rehearsal protocol, in which participants are instructed to say any word that comes to mind out loud while studying the list items. Seamon et al found that recall and recognition of the critical lure was high, but the level of false memory did not differ between the overt-rehearsal and a silent-rehearsal group. For the overt-rehearsal group, the critical lure was indeed overtly rehearsed and spontaneous mention of the lure during study enhanced false recall at test, but contrary to the expectations from Underwood's IAR-account, false recall still occurred for a large proportion of lures that were not overtly rehearsed. In addition, the level of false recognition was the same for lures that were and lures that were not rehearsed. These results have been interpreted as providing evidence against the necessity that the IAR leads to conscious awareness of the lure and instead support theories that assume an unconscious automatic generation of the lure (Roediger, Balota & Watson, 2001) or gist trace (Brainerd & Reyna, 2002).

However, in a recent article, Laming (2006) investigated the overt-rehearsal protocol and concluded that the processes underlying rehearsing out loud during list presentation and the processes underlying memory retrieval are the same. This has important implications for the interpretation of the Seamon et al data, which can be re-interpreted as showing that the retrieval of the lure increases the probability that it will be recalled again (a basic memory recall effect), but does not increase the probability of recognizing the lure at a later stage. This latter aspect is particularly interesting given its possible meaning that falsely recalling an item does not lead to increments in the memory trace.

Despite this alternative interpretation there is other evidence suggesting a locus at encoding and not at retrieval. For example, the amount of false memories increases with the number of related words (Robinson & Roediger, 1997) and the number of related words presented prior to the lure word in a test list has no effect on the false alarm rate (Marsh, McDermott & Roediger, 2004).

The above findings are complicated by the fact that the memory paradigms tap into both the episodic and semantic systems, which would not be a problem if it were not the case that episodic encoding is influenced by the semantic structure of the list (e.g., Davelaar, et al., 2006) and that retrieval of items is determined jointly by episodic traces and semantic associations (e.g., Davelaar, et al., 2006; Howard & Kahana, 2001). In what follows next, I will review a recent memory theory that could inform a new way of analyzing the DRM data.

3. Contextual retrieval

3.1. *Context models*

In the memory literature there is a re-interest in computational models of free recall memory (Davelaar, et al., 2005; Howard & Kahana, 2002). This is partly due to insights gained from novel analytical techniques (Kahana, 1996) and partly due to the resurrection of the debate on the existence of a short-term buffer (Davelaar, et al., 2005; Neath & Brown, 2006). Despite the critical differences between the models, all incorporate a context system. This system is best described of consisting of features or elements that are either active or not. This distributed context system changes during the course of list presentation, such that with the presentation of every new item some active features become inactive and some inactive features become active. The novel aspect of this incarnation of a classic Markov system which has been used by numerous memory theorists as early as with Estes (1955) is that the trigger for changing the state of the contextual features depends on the item that is presented (Howard & Kahana, 2002). In the temporal context model (TCM: Howard & Kahana, 2002), the contextual change is governed by the retrieval of pre-experimental context of the presented item.

Howard and Kahana (1999; 2002) assume that every item resides in the long-term memory system and is associated with its unique context that becomes activated when the item is encountered. The ongoing experimental context is then combined with the retrieved pre-experimental context and associated with the item. The resulting state in the context system is then combined with the retrieved pre-experimental context of the next item and so on. Memory performance is a function of the similarity between the context state at test and during encoding. This by itself is no different than a random-changing context model. However, Howard and Kahana (2002) went further and assumed that during free recall, the pre-experimental context of the *just-reported* item is retrieved and used to retrieve the next item. This aspect sets TCM apart from other context models and allows it to capture the asymmetry in conditional response probabilities.

Kahana (1996) showed that in a free recall paradigm two items that are reported consecutively during the retrieval phase tend to have been presented close together during encoding. Specifically, the probability of retrieving item j given item i is a decreasing exponential function of the distance, $|i-j|$, between the two items in the presented list. This pattern is readily explained in random-changing context model in that the encoded context of items presented in close temporal distance is similar that items with larger distance. With the assumption of retrieval of encoded context, conditional response probabilities (CRP) will reflect this contextual overlap. Kahana (1996) also showed that the CRP-functions are asymmetric and biased in the forward direction. In other words given item n the next item to be retrieved is more likely to be n+1 than n-1 even though both are more likely than n+2 or n-2. Howard and Kahana (2002) showed that this asymmetry can be accounted for by assuming retrieval of pre-experimental context. During encoding, item n triggers the retrieval of its pre-experimental context which gets combined in the ongoing context. Therefore, item n+1 gets encoded with a contextual state that is a combination of its retrieved pre-experimental context, some pre-experimental context of item n and so on. Importantly, item n-1 has not been encoded with any contextual features from n's pre-experimental context. In other words, the forward asymmetry is already present in the encoded context. During the retrieval phase one has only to assume that the pre-experimental context is retrieved after the retrieval of the item. The combined context will then favor item n+1 given retrieval of item n.

The asymmetry is a direct result of encoding the presented item and retrieving its specific (i.e., pre-experimental) features. Therefore any item that came to mind during encoding left its mark in the ongoing context and has obtained new episodic associations with the item that preceded it. This forms the basis of the proposed contextual retrieval analysis, which I will apply to the DRM paradigm.

3.2. *Contextual retrieval analysis*

As discussed above, when one assumes TCM to provide an accurate account of memory retrieval processes, one can build on this and develop a theoretically-motivated analysis. Here, I will use the insights gained from TCM to investigate whether the critical lure in a free recall task of the DRM paradigm was encoded in the ongoing context or was produced during retrieval only. In other words, the analysis focuses on the probability of reporting the lure given retrieval of item i (predecessors; CRP-pre) and the probability of reporting item j given report of the lure (successors; CRP-suc).

I assume that the critical lure has been encoded if lure-specific features have been activated and encoded in the ongoing context. This seems trivial, but it means that I assume that the lure will therefore have a verbatim trace. Specifically, the CRP-pre will be peaked on n, where n is the item *after* which the lure came to mind and got associated with the ongoing context. Of course, when the probability of the lure coming to mind is uniformly distributed over all n, the CRP-pre is indistinguishable from the scenario where the lure is only retrieved through the use of the gist trace.

Assuming that the lure was activated during encoding, it could trigger the retrieval of its pre-experimental context and perhaps again during retrieval. If this happened, the CRP-suc should be a peaked function centered on n, where n is the item *before* which the lure came to mind. When the lure did not trigger pre-experimental context during encoding or retrieval, this peaked function will not be observed. It is possible that the coming to mind of the lure during encoding or retrieval is not accompanied by retrieval of the lure's pre-experimental context. In those cases the ongoing context will not be affected.

4. Experiments

4.1. *Experimental method*

The contextual retrieval analysis was applied to results of two experiments. The first experiment aimed at investigating the relation between working memory capacity and false memory (cf. Watson, et al., 2005). The second set of results forms a subset of a larger study on memory capacity and executive function. Although the initial reason for conducting the studies varied, the analyses attempted here are meant to be applicable to any past and future datasets obtained from a DRM free recall paradigm.

In experiment 1, 29 participants (18 female; mean age 39 years) were tested on 37 lists of 15 words each for immediate free recall. The first trial was a practice trial to familiarize the participant with the procedure. The remaining 36 trials were DRM-lists taken from the Roediger et al (2001). Each list of was presented visually at a rate of one word per second. In experiment 2, 70 participants (40 female; mean age 29 years) were tested on 2 practice trials and 10 DRM lists which were presented visually at a rate of one word per second. Both experiments included the operation span task (Turner & Engle, 1989) as a measure of working memory capacity. Although unrelated to the present analysis, it is noteworthy to mention that participants in the second experiment had a higher span score than participants in the first experiment (M_{second} = 21.4 [sd = 9.7] vs. M_{first} = 16.5 [sd = 9.4]; $t(97) = 2.29$, $p < .05$).

The contextual retrieval analysis, i.e., the consideration of CRP-pre and CRP-suc, requires that the lure is reported and that it was preceded and succeeded by list items. Given that the false alarm rate tends to be around 30%, the analysis necessarily has to ignore individual differences and treat all lists across all participants as coming from a single participant. This problem currently excludes the use of inferential statistics and will therefore form the basis of future developments of this approach.

4.2. *Results*

Out of 1044 trials (29 participants x 36 DRM-lists) in experiment 1, there were a total of 319 false memories, out of which 205 were preceded and succeeded by words from the list. For experiment 2, out of 700 (70 participants x 10 DRM-lists), there were a total of 271 false memories of which 201 were preceded and succeeded by words from the list.

For comparison with Watson et al (2005), a separate analysis was conducted on the relation between working memory and false memory. Using a median-split on the data (excluding the person with the highest span score), it was found that high span score was associated with high veridical recall ($M_{high\ span}$ = .94 [sd = .04], $M_{low\ span}$ = .90 [sd = .05]; $t(26) = 2.08$, $p < .05$), but not with false recall ($M_{high\ span}$ = .26 [sd = .16], $M_{low\ span}$ = .36 [sd = .16]; $t(26) = 1.67$, $p = .107$). This null-effect is likely to be due to low numbers of people tested, despite the large probability of false memory.

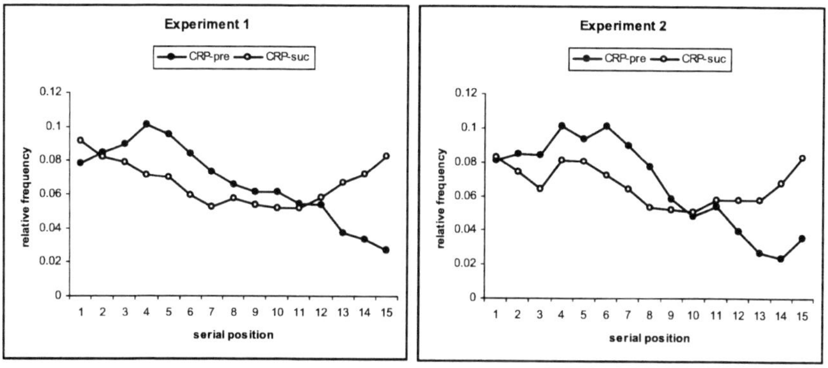

Figure 1: Relative frequency distributions of CRP-pre and CRP-suc from experiments 1 and 2.

Figure 1 presents the results for CRP-pre and CRP-suc for the two experiments. The CRPs were calculated by tabulating the number of occurrences that item n *immediately* preceded (succeeded) the lure in the verbal recall report and dividing this by the total number of times that item n preceded (succeeded) the lure at all (but not necessarily immediately). This procedure controls for the fact that in immediate free recall, participants tend to start with recall of the last presented items which tend to precede the report of the lure. In order to allow for visual comparison across the experiments, the resulting frequency distributions were normalized. Given the novelty of the present approach it is important that the results are the same despite across-experiment differences in number of participants and DRM-lists.

The CRP-pre shows a peaked distribution with a peak around serial position 4-6. As discussed above, we would expect a peaked distribution if (1) the lure word came to mind after item n, (2) the lure word (including lure-specific features) got encoded in the ongoing context, and (3) retrieval of item n causes reinstatement of the context that was present at encoding. The correlation between the two experiments was $r(15) = .773$, $p < .001$, indicating strong consistency.

The CRP-suc does not show a similar peaked distribution and instead resembles a standard U-shaped serial position function. As discussed above, we would expect a peaked distribution if (1) the lure word came to mind before item n, and (2) the lure word triggered the retrieval of pre-experimental context during encoding and (3) during retrieval. The correlation between the two experiments was sizable, $r(15) = .426$, but not significant $p = .113$. This is primarily due to the "hump" at positions 4-6 in experiment 2. Closer scrutiny of the data showed that this "hump" was caused by a subset of the 10 trials, with some words being more memorable across participants, suggesting that a random assignment of items to positions is desirable.

4.3. *Interpretation*

The main purpose of this paper is to introduce and demonstrate the utility of a type of analysis that is informed by recent computational theorizing about memory recall. The pattern of data presented in Figure 1 supports the following conclusions. First, the peaked distribution of CRP-pre implies (within the TCM-informed analysis) that the critical lure comes to mind after encoding about 4 items (the location of the peak). When this happens, lure-specific features are activated too and are encoded in the ongoing episodic context. At retrieval, when item 4 is retrieved, it triggers contextual retrieval of its pre-experimental context and thus will favor the retrieval of the lure. This conclusion does not depend on an overt-rehearsal task (see Seamon, et al., 2002), which can be criticized as being a retrieval-during-encoding procedure. In addition, the mere fact that a peak is observed implies that the lure word has a verbatim trace. This is a highly controversial statement (but see for similar implications, Kimball, Smith & Kahana, 2007) and favors memory-based accounts over decision-based accounts.

Second, the lack of a peaked distribution for CRP-suc suggests that at least one of the afore-mentioned assumptions has not been met. One parsimonious explanation could be that although the CRP-pre shows that the lure word came to mind during encoding, the lure word in itself did not trigger the retrieval of pre-experimental context (during encoding or retrieval). In other words, an imagined word or event can get encoded in ongoing context, but it will not lead to change of that context. The U-shaped distribution of the CRP-suc merely reflects the underlying structure of contextual retrieval from item n to item n+1 in the list (due to space-limitations a full analysis of this part is not possible).

4.4. *Methodological considerations*

The two experiments presented here use immediate free recall. Although this is an often-used test in the memory literature it has some drawbacks. First, the last few items in the list are retrieved using short-term memory and therefore retrieval of those items is less reliant on episodic associations and may enhance contributions from semantic associations (Davelaar, et al., 2006). Second, it is known that the lure word is retrieved later in the recall protocol (Roediger & McDermott, 1995) and therefore the resulting distribution of CRP-pre may become artificially biased to have low values for later serial positions. An alternative is to use a delayed free recall task, in which after the final item is presented the participant engages in a demanding distractor task that is aimed at displacing the items from the short-term buffer.

The two experiments also used a specific order of the list items, as used by many researchers. List items are presented in decreasing order of associative strength to the lure word. Although in itself this is not a problem, it does beg the question whether the peak for the CRP-pre is at earlier serial positions because of the stronger associates or because of generally increased associative activations (cf. Robinson & Roediger, 1997). In addition to avoid list-specific artefacts in the distribution, random allocation of items to serial positions is desired.

These considerations generally apply when using memory recall, but may also apply in recognition experiments. Recently, Schwartz et al (2005) showed that temporal context effects are also observed in recognition tasks (although the asymmetry is absent).

5. Discussion

In this paper, I have introduced a new analytical technique that can be used in addressing the origin of false memories in free recall tasks. I made the critical assumption that in order to claim that the earliest cognitive operation that can lead to a false memory lies at encoding, lure-specific features should have been activated and encoded in the ongoing context. This necessarily means that I assume then that the lure has a verbatim trace. If a verbatim trace exists the memory is susceptible to retrieval processes that operate on those traces. I appealed to the temporal context model by Howard and Kahana (2002) to suggest an analysis of conditional probabilities. The CRP-pre is the conditional response probability of reporting item n given report of the lure next. This CRP-distribution is peaked over item n in the list after which the lure came to mind (and subsequently was encoded). The CRP-suc is the conditional response probability of reporting item n immediately after report of the lure word. This CRP-distribution is peaked over item n in the list before which the lure came to mind, retrieved its pre-experimental context during encoding which was added in the ongoing context, and again during retrieval.

I presented reanalyses of two experiments using immediate free recall and showed results that are consistent with the interpretation that the non-presented lure word came to mind during encoding of associated items (i.e., the lure word has a verbatim trace), but the lure word did not trigger retrieval of its pre-experimental context. Further studies (and reanalyses of existing studies) are needed to test this interpretation and validate this new analytical tool, which I have referred here to as "contextual retrieval analysis". The analysis depends on the TCM-assumption that processed items retrieve pre-experimental context, which then gets combined in ongoing context. One could certainly argue against this assumption and as of writing no independent test of this assumption has been reported in the memory literature.

In this paper, I used insights gained from a recent complex computational model of free recall memory and applied them in an analysis of empirical data that have yet to be modeled. This analysis can be conducted by any empirical researchers without the need of using a computational model or understanding its underlying mathematics. In the future, more computational modelers may use insights gained from their work in the development of new analytical tools in the same way as statistical tools became more mainstream.

Acknowledgments

I thank Hannah Dickson, Violeta Dobreva, and Helge Gillmeister for data collection. The data of experiment 2 was collected in a study funded by The British Academy (SG-38634).

References

Brainerd, C. J., & Reyna, V. F. (2002). Fuzzy trace theory and false memory. *Current Directions in Psychological Science, 11*, 164-169.

Davelaar, E. J., Goshen-Gottstein, Y., Ashkenazi, A., Haarmann, H. J., & Usher, M. (2005). The demise of short-term memory revisited: empirical and computational investigations of recency effects. *Psychological Review, 112*, 3-42.

Davelaar, E. J., Haarmann, H. J., Goshen-Gottstein, Y., & Usher, M. (2006). Semantic similarity dissociates short- from long-term recency: testing a neurocomputational model of list memory. *Memory & Cognition, 34*, 323-334.

Gallo, D. A. (2006). *Associative illusions of memory: false memory research in DRM and related tasks.* New York: Psychology Press.

Howard, M. W., & Kahana, M. J. (1999). Contextual variability and serial position effects in free recall. *Journal of Experimental Psychology: Learning, Memory, and Cognition, 25*, 923-941.

Howard, M. W., & Kahana, M. J. (2001). When does semantic similarity help episodic retrieval? *Journal of Memory and Language, 46*, 85-98.

Howard, M. W., & Kahana, M. J. (2002). A distributed representation of temporal context. *Journal of Mathematical Psychology, 46*, 269-299.

Kahana, M. J. (1996). Associative retrieval processes in free recall. *Memory & Cognition, 24*, 103-109.

Kimball, D. R., Smith, T. A., & Kahana, M. J. (2007). The fSAM model of false recall. Psychological Review, 114, 954-993.

Laming, D. (2006). Predicting free recalls. *Journal of Experimental Psychology: Learning, Memory, and Cognition, 32*, 1146-1163.

Marsh, E. J., McDermott, K. B., & Roediger, H. L. (2004). Does test-induced priming play a role in the creation of false memories? *Memory, 12*, 44-55.

Miller, M. B., & Wolford, G. L. (1999). Theoretical commentary: the role of criterion shift in false memory. *Psychological Review, 106*, 398-405.

Robinson, K. J., & Roediger, H. L. (1997). Associative processes in false recall and false recognition. *Psychological Science, 8*, 231-237.

Roediger, H. L., & McDermott, K. B. (1995). Creating false memories: remembering words not presented in lists. *Journal of Experimental Psychology: Learning, Memory, and Cognition, 21*, 803-814.

Roediger, H. L., Balota, D. A., & Watson, J. M. (2001). Spreading activation and the arousal of false memories. In H. L. Roediger, J. S. Nairne, I. Neath, & A. M. Surprenant (Eds.), *The nature of remembering: essays in honor of Robert G. Crowder* (pp. 95-115). Washington, DC: American Psychological Assocation.

Roediger, H. L., Watson, J. M., McDermott, K. B., & Gallo, D. A. (2001). Factors that determine false recall: a multiple regression analysis. *Psychonomic Bulletin & Review, 8*, 385-407.

Seamon, J. G., Lee, I. A., Toner, S. K., Wheeler, R. H., Goodkind, M. S., & Birch, A. D. (2002). Thinking of critical words during the study is unnecessary for false memory in the Deese, Roediger, and McDermott procedure. *Psychological Science, 13*, 526-531.

Watson, J. M., Bunting, M. F., Poole, B. J., & Conway, A. R. A. (2005). Individual differences in susceptibility to false memory in the Deese-Roediger-McDermott paradigm. *Journal of Experimental Psychology: Learning, Memory, and Cognition, 31*, 76-85.

ANOTHER REASON WHY WE SHOULD LOOK AFTER OUR CHILDREN

JOHN A. BULLINARIA

School of Computer Science, The University of Birmingham
Edgbaston, Birmingham, B15 2TT, UK
j.bullinaria@physics.org

In many ways, it seems obvious that we should look after, feed and protect our children. However, infants of some species are expected to look after themselves from a very early age. In this paper, I shall present a series of simulations that explore the hypothesis that neural network learning issues alone are sufficient to result in the evolution of long protection periods. By evolving populations of neural network systems that must learn to perform simple classification tasks, I show that lengthy protection periods will emerge automatically, despite associated costs to the parents and children.

1. Introduction

Most humans accept that it is part of their role as parents to look after their children until they are old enough to fend for themselves, and it is clear that the children would have an extremely low survival rate if that did not happen. But why have humans evolved to be like that?

Many species are *precocial*, with young that are born well developed and requiring very little parental care. Others are *altricial*, with relatively helpless young requiring periods of parental care before they are able to survive on their own. Human infants are particularly altricial, even compared with other primates, requiring extended periods of parental protection and support (e.g., Lamb, Bornstein & Teti, 2002). For altricial species there are usually two important processes happening during the protection stage – the infants are growing, and they are learning. Human infants do require a lot of growing after birth, and parental protection does provide obvious advantages in term of survival. But why such extended periods compared with other primates? The need to learn will depend on how much innate knowledge the individuals are born with. It is likely that learning is crucial when relatively complex behaviour is required, or when the properties of the environment are variable and each new-born infant needs to learn to adapt accordingly. Individuals will also need to learn to adapt their control processes to compensate for changes caused by their growing (Bullinaria, 2003a). All these processes are clearly applicable to humans. It seems, however, that humans do have excessively long protection periods, and many parents may wonder if they really do need to look after their children for quite so long. Their children might also wonder if they wouldn't be better off "leaving home" and embarking on their reproductive careers at an earlier age.

In this paper, I explore, through a series of simulations, one reason why evolution might have favoured long protection phases in humans. There are actually many possible reasons (e.g., see Sloman & Chappell, 2005), but here I shall focus on the hypothesis that learning issues alone are sufficient to result in the evolution of long protection periods. Moreover, since most human learning takes place in their brains, it is neural network learning that will be studied. I have previously run simulations of the evolution of populations in which learning individuals of all ages compete for survival according to their performance on simplified classification tasks. Not surprisingly, individuals evolved that not only learned how to perform well, but were also able to learn *quickly* how to achieve that good performance. However, it was also observed that the pressure to learn quickly could also have the unfortunate side effect of leading to risky learning strategies that sometimes resulted in very poor performance (Bullinaria, 2007).

In this study, I shall again present results from evolved neural network systems that must learn to perform well on simple classification tasks. One might consider reducing the computational resource requirements of the simulations by using a learning mechanism, or an approximation of learning, that is simpler than an artificial neural network. The problem with attempting this is that the error distributions and associated fitness levels during learning depend in a complex manner on the learning algorithm and its evolved parameters, and these in turn depend in a non-trivial way on the evolutionary pressures and population age distributions which are affected by the protection period we are attempting to study. It is almost impossible to predict what distributions of all these things will emerge across the evolving populations. With so many unknowns and complex interactions, the only reliable way to proceed in the first instance is to run the full evolutionary neural network simulations. Future studies will then be able to safely abstract out the key features for exploration of further issues.

The remainder of this paper will show that evolved neural network systems do exhibit better adult performance if protection from competition is provided during the children's early years. Moreover, if the length of the protection period is allowed to evolve, it does result in the emergence of relatively long protection periods, even if there are other costs involved, such as the children not being allowed to reproduce during their protection phase, and the parents suffering increased risk of dying while protecting their young.

2. Evolving Neural Network Systems

The idea here is to mimic the crucial features of the evolution of most animal populations, but concentrate on the aspects of fitness associated with neural network learning. We therefore take a whole population of individual neural networks, each specified by a set of innate parameters, and expect them to learn from a continuous stream of input patterns how to classify future input patterns. Those inputs could, for example, correspond to specific features of other animals, and the desired output classes could correspond to being dangerous, edible, and such like. Each individual then has a fitness measure determined by how well it classified each new input *before* discovering (somehow) its correct class and learning from it. If the individuals compete to survive and procreate according to their relative fitnesses, we can expect populations of increasing fitness to emerge.

To proceed empirically, we need to concentrate on a specific concrete system, and it makes sense to follow one that has already proved instructive in the past (Bullinaria, 2007). Real-world classification tasks typically involve learning non-linear classification boundaries in a space of real valued inputs. Taking the set of classification tasks corresponding to two dimensional continuous input spaces with circular classification boundaries proves simple enough to allow extensive simulations, yet involves the crucial features and difficulties of real world problems. Each "new-born" neural network is assigned a random classification boundary which it must learn from a stream of randomly drawn data points from the input space, that we can take to be normalized to a unit square. The natural performance measure we shall use is the generalization ability, i.e. the average number of correct outputs (e.g., output neuron activations within 0.2 of their binary targets) before training on them.

We shall take our neural networks to be traditional fully connected Multi-Layer Perceptrons with one hidden layer, sigmoidal processing units, trained by gradient descent using the cross-entropy error function. As previous studies have shown (Bullinaria, 2003b), one gets better performance by evolving separate learning rates η_L and initial weight distributions $[-r_L, +r_L]$ for each of the four distinct network components L (the input to hidden weights IH, the hidden unit biases HB, the hidden to output weights HO, and the output unit biases *OB*), rather than having the same parameters across the whole network. These, together with the standard momentum parameter α and weight decay regularization parameter λ, result in ten evolvable innate parameters for each network. It is also possible to evolve the number of hidden units, but since the evolution invariably results in the networks using the maximum number we allow, slowing down the simulations considerably, we keep this fixed at 20 for all networks, which is more than enough for learning the given tasks.

For the simulated evolution, we need a single unit of time that covers all aspects of the problem, so we define a "simulated year of experience" to be 1200 training data samples, and compute the fitness of each individual at the end of each year as an average over that year. This number ensures that each individual has its performance sampled a reasonable number of times during its learning phase. Then, using random pair-wise fitness comparisons (a.k.a. tournaments) at the end of each year, we select up to 10% of the least fit individuals to be killed by competitors and removed from the population. In addition, to prevent the populations being dominated by a few very old and very fit individuals, a random 20% of individuals aged over 30 simulated years die of old age each year and are removed from the population.

A fixed population size of 200 is maintained throughout (consistent with the idea that there are fixed total food resources available to support the population), with the removed individuals being replaced by children generated from random pairs of the most fit individuals. Each child inherits innate parameters that are chosen randomly from the corresponding range spanned by its two parents, plus a random mutation (from a Gaussian distribution) that gives it a reasonable chance of falling outside that range. These children are protected by their parents until they reach a certain age and cannot be killed by competitors before then. The direct cost to the children is that they are not allowed to have any children of their own before they leave the protection of their parents. The implicit cost to the parents is that, the more the children are protected, the higher the chance they stand of being in the 10% of the population that are killed each year. In practice, a protected child might receive more from its parents than simple protection (e.g., teaching as well), but throughout these simulations we avoid all potential confounds by always using the set-up that is *least* likely to lead to a positive learning effect.

Various aspects of this basic evolving neural network idea have already been explored in some detail elsewhere (Bullinaria, 2001, 2003a, 2003b, 2007). The two crucial new issues to be investigated here are:

1. How does protecting the children affect the individuals' performance?

2. If the duration of the protection period is free to evolve, what happens?

The first question can be conveniently explored by fixing the protection period at a number of different values by hand, evolving the other innate parameters as before, and comparing the levels of performance that emerge. The second can be answered by running similar simulations with the protection period allowed to be an additional evolvable parameter, and analyzing what happens. The various simulation results are presented in the next three sections.

3. Simulation Results for Different Protection Periods

The natural starting point is to run the evolutionary neural network simulations with a few carefully selected fixed protection periods to determine if that makes any difference to the evolved populations. Since the individuals typically learn their given task in 10 to 20 simulated years, and start dying of old age at age 30, it makes sense to begin by looking at protection periods of 1, 10 and 20 years. Figure 1 shows the evolution of the learning rates for these three cases, with means and variances over six runs. The evolved parameters and low variances across runs are similar to those found in previous studies (Bullinaria, 2007).

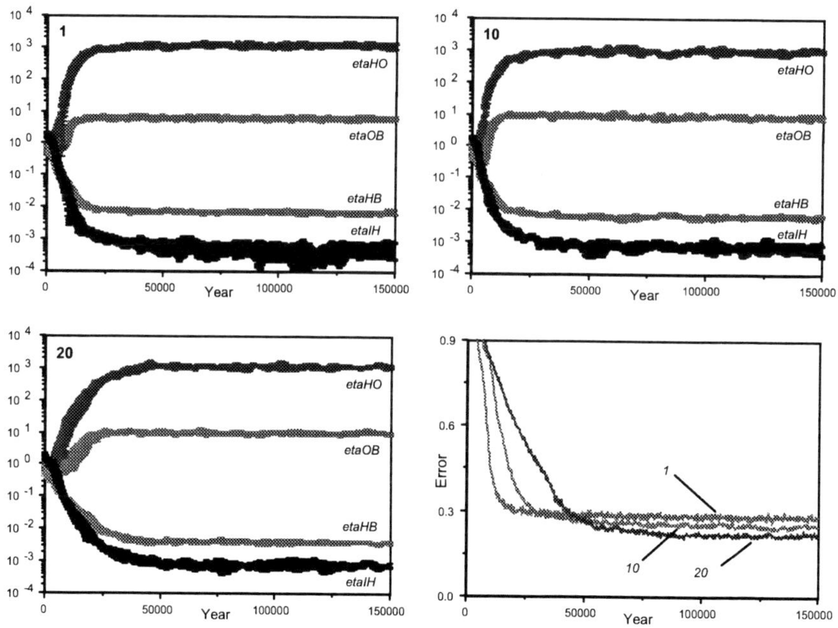

Figure 1: Evolution of the learning rates for protection periods of 1, 10 and 20 years, and comparison of the evolution of the corresponding mean error rates.

The evolving parameters in each case have settled down by about 50,000 years, and subtle differences can be seen between the final values.

The final panel in Figure 1 compares the generalization performance means across populations during evolution for each protection period. It shows that the evolutionary process is much slower to settle down for the longer protection periods, but longer protection periods *do* appear to have an advantage in terms of final evolved performance. However, these population means hide complex age dependent error distributions, and the population age distributions are unlikely to be the same across the various cases, so to see whether there is a real evolutionary advantage of increased protection periods, we need to simulate their evolution. This is done in the next section.

4. Allowing the Protection Period to Evolve

If the protection period is allowed to evolve, the evolution of that period and the associated learning rates that emerge are as shown in Figure 2, again with means and variances across six runs. There is higher variance in all the parameters, compared to the fixed protection period runs, until the protection period has settled down after about 40,000 years. Early on in evolution, when the populations are relatively poorly performing, the protection period rises rapidly to about 25 years, but then falls and settles to around 16 years.

Comparison of the averages and variances of the crucial evolved population properties are presented in Figure 3, for both the fixed and evolved protection periods. As one would expect, the number of deaths per year due to competition decreases, from the maximum of 20 per year, as the protection period increases, and this inevitably increases the average age of the population. In turn, more individuals survive to old age, and consequently the deaths per year due to old age increase slightly. Overall, there is still a net reduction in deaths per year,

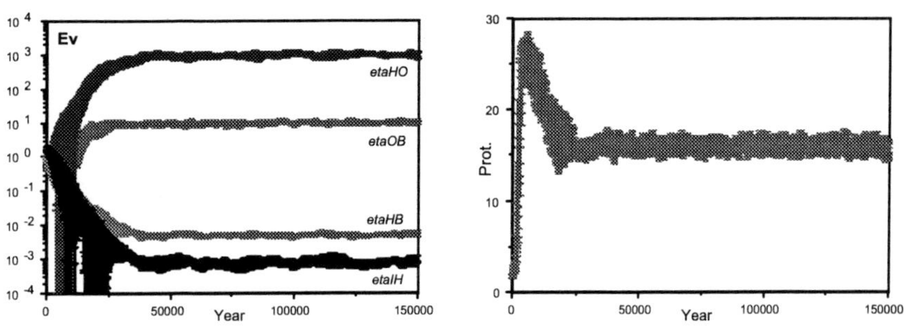

Figure 2: Evolution of the learning rates when the protection period is allowed to evolve, and the evolution of the protection period.

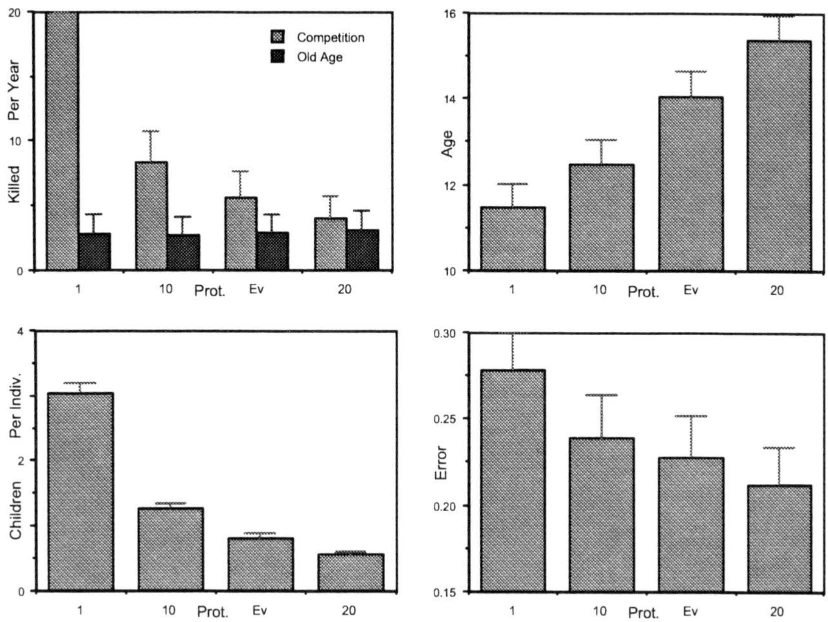

Figure 3: Comparison of the evolved population averages and variances for the various fixed (*1*, *10*, *20*) and evolved (*Ev*) protection periods: deaths per year, ages, children per individual, and performance error rates.

and so, given the fixed population size, the average number of children per individual at any given time decreases with the protection period. (Note that this is independent of any direct introduced cost of parents protecting more children.) Finally, the average population performance error rate (i.e. inverse fitness) is seen to fall steadily with increasing protection periods. All these trends vary monotonically with protection period, and the evolved protection period population results are consistent with what would be expected from their evolved period of 16 years.

The obvious next question is, given that the average population fitness increases with protection period, why is it that the evolved protection age does not end up higher than 16? Actually, the distributions in Figure 3 already provide us with some clues. First, given that the older individuals will have had more time to learn, they will inevitably be fitter by our criteria, and hence the increases in average age will automatically translate into increased population fitness, even if each individual were no better as a result of the protection period. Moreover, even if there were individual fitness advantages, the reduced number of children per individual for increased protection periods will place

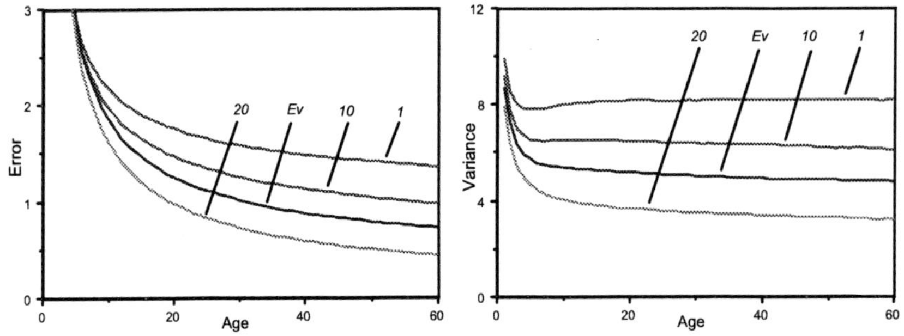

Figure 4: Mean errors and variances during learning for evolved individuals, with evolved protection period (*Ev*) and the three fixed protection periods (*1*, *10*, *20*).

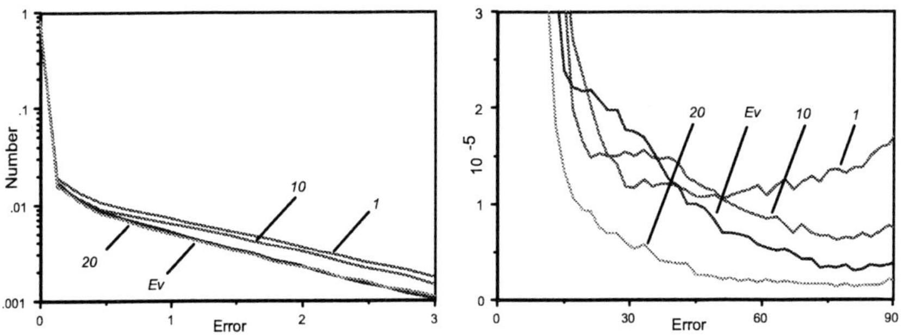

Figure 5: The peaks and tails of the error distributions for evolved individuals aged between 50 and 60, for each of the four protection period cases (*Ev*, *1*, *10*, *20*).

individuals with long protection periods at an evolutionary disadvantage, and this will tend to decrease the evolved protection periods. To understand the advantages and disadvantages to individuals, and explore the detailed effects of such trade-offs, we need to look more carefully at the individual fitness profiles. This will be done in the next section.

5. Analysis of the Evolved Performance

The means and variances of the individual error rates (i.e. inverse fitness) during learning are shown in Figure 4, and there do indeed appear to be significant advantages for protracted protection periods. However, the error distributions for this type of problem are known to be rather skewed, with the residual mean errors due largely to instances from the long tails of very large errors (Bullinaria, 2007). This is clear in the peaks and tails of the error distribution for individuals aged between 50 and 60 years shown in Figure 5. There is a

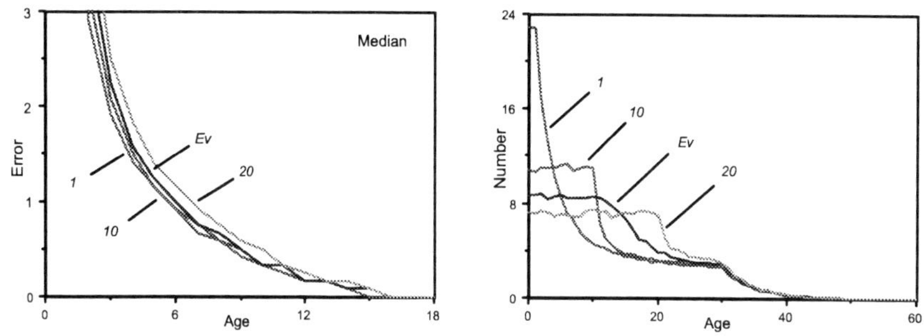

Figure 6: The median error rates during learning and the age distributions of the evolved populations, for each of the four protection period cases (*Ev*, *1*, *10*, *20*).

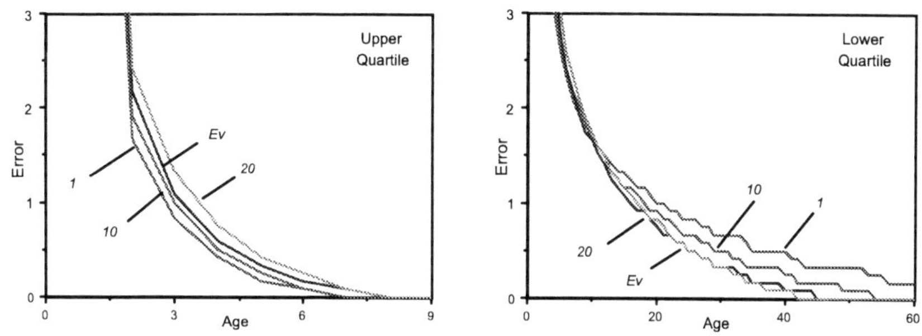

Figure 7: The upper and lower quartile error rates during learning for evolved individuals, for each of the four protection period cases (*Ev*, *1*, *10*, *20*).

massive peak around zero errors, as one would expect at that age, but there also remain significant numbers of very large errors. This is a common feature of evolutionary processes that encourage fast learning (Bullinaria, 2007), and longer protection periods, which limit the need for fast learning at early ages, seem to alleviate the problem.

One can get a better idea of the population performances, that is not skewed by a few instances of very poor performance, by looking at the medians rather than the means. The median error rates during learning, shown on the left of Figure 6, are in accordance with the expectation of learning the task essentially perfectly by a certain age, but there is surprisingly little difference in median performance across the four protection periods. There is at most two years difference in learning across the cases, despite the twenty years range of protection periods, and the wide variations in the age distributions of the evolved populations shown on the right of Figure 6. Each age distribution is fairly flat during the protection period, then falls off due to competition till the individuals start dying of old age from the age of 30, at which point there is an exponential fall to zero. The upper and lower quartile error rates are shown in Figure 7. The faster learning quartiles are still remarkably similar to each other, with just a slight increase in learning speed resulting from shorter protection periods. However, there are clear differences in the slower learning quartiles, with large improvements seen for the longer protection periods, as already evident in Figures 4 and 5.

6. Discussion and Conclusions

The above simulations have established that longer protection periods *do* offer a clear learning advantage, and relatively little disadvantage. So we now need to return to the question of what it is that prevents the evolved protection period from becoming even longer. But first, we need to make sure that the evolved period we found is not simply an artifact of the chosen evolutionary process. This can be checked by freeing the protection period in each of the three fixed period evolved populations, and allowing them to evolve further. The results of this are shown on the left of Figure 8, with means and variances across six runs. In each case there is either a rise or fall to the same evolved period of around 16 years that emerged before.

A second check involves combining the evolved populations from the four cases into one big population, and then allowing natural selection to take its course. Since each case has already been optimized by evolution, no further crossover and mutation was allowed. The outcome is shown on the right of Figure 8, with means and variances across twelve runs. There is quite a large variance across runs, but individuals with the evolved protection period consistently come to dominate the whole population. Individuals with virtually no protection are wiped out almost immediately.

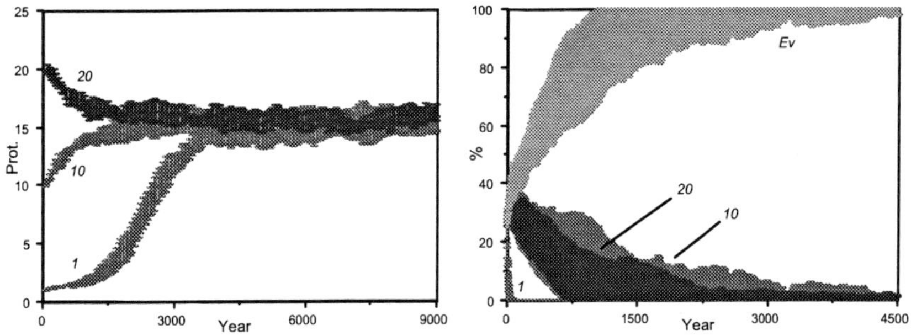

Figure 8: Evolutionary improvement of protection periods (left) and competition between the evolved populations from the four protection period cases (right).

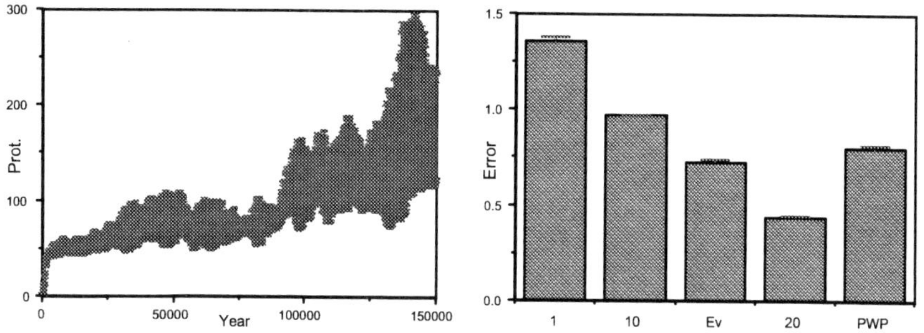

Figure 9: Evolution of the protection period when procreation while protected (PWP) is allowed (left), and comparison of the mean error rates at age 60 (right).

The natural conclusion from the results in Figure 8 is that, although there are clear learning advantages to having longer protection periods, the best periods from an evolutionary point of view are shorter than they could be. This can be understood in terms of the number of children per individual seen in Figure 3. Because the individuals effectively have fixed life-spans, extended periods of protection will use up a significant proportion of the potential procreation period and thus put those populations at a serious evolutionary disadvantage. The evolutionary simulations are able to establish a suitable trade-off value for the protection period, that balances the improved performance against the loss of reproductive opportunities.

The obvious check of this conclusion is to repeat the whole evolutionary process with procreation allowed while being protected. As seen on the left of Figure 9, the protection period then evolves to be way beyond the normal life-span of the individuals, so that there are no deaths at all due to competition, only due to old age. What this scenario has re-introduced, however, is the need to compete at all ages to procreate, and this encourages faster learning again. Of course, that brings back with it the unwanted associated side effects, such as the use of risky learning strategies that sometimes result in persistent very poor performance at all ages. This can be seen in the increased mean error rates at age 60 shown on the right of Figure 9. One could imagine allowing individuals to procreate randomly without having to compete to do so, and that would remove the pressure to learn quickly, but that would leave no evolutionary pressure to improve fitness at all, and the individual performances would end up even worse. It seems then, that there is a real advantage to preventing offspring from reproducing while being protected, and that goes beyond enhancing the parents' own reproductive success rate.

In conclusion, the results presented in this paper have shown how evolutionary neural network simulations can begin to address aspects of human behaviour such as the protection of offspring. Of course, there are many related issues that remain to be taken into account in future work. First, more attention could be paid to the changes in the learning experience that might result from the parental protection, for example due to guided exploration, exploration without risk, "teaching", and so on. The costs to parents of protecting their children should also be accounted for more carefully, particularly for situations where each parent has to protect many children for long periods. We also need to consider the evolutionary pressures and consequences that would arise due to the introduction of competition with, and co-evolution with, other species. The evolved protection periods are also likely to interact with changes to the natural life-span of the species, and that needs to be explored. There are also many other "life history" factors which may evolve (e.g., Stearns, 1992; Roff, 2002), with associated trade-offs, and these could also usefully be incorporated into improved future models.

References

Bullinaria, J.A. (2001). Simulating the Evolution of Modular Neural Systems. In: *Proceedings of the Twenty-Third Annual Conference of the Cognitive Science Society*, 146-151. Mahwah, NJ: Lawrence Erlbaum Associates.

Bullinaria, J.A. (2003a). From Biological Models to the Evolution of Robot Control Systems. *Philosophical Transactions of the Royal Society of London* A, **361**, 2145-2164.

Bullinaria, J.A. (2003b). Evolving Efficient Learning Algorithms for Binary Mappings. *Neural Networks*, **16**, 793-800.

Bullinaria, J.A. (2007). Using Evolution to Improve Neural Network Learning: Pitfalls and Solutions. *Neural Computing & Applications*, **16**, 209-226.

Lamb, M.E., Bornstein, M.H. & Teti, D.M. (2002). *Development in Infancy: An Introduction*. Mahwah, NJ: Lawrence Erlbaum Associates.

Roff, D.A. (2002). *Life History Evolution*. Sunderland, MA: Sinauer Associates.

Sloman, A. & Chappell, J. (2005). The Altricial-Precocial Spectrum for Robots. In *Proceedings of the International Joint Conference on Artificial Intelligence*, 1187-1193. IJCAI.

Stearns, S.C. (1992). *The Evolution of Life Histories*. Oxford, UK: Oxford University Press.

Section II

Language

A MULTIMODAL MODEL OF EARLY CHILD LANGUAGE ACQUISITION

ABEL NYAMAPFENE

School of Engineering, Computing and Mathematics, University of Exeter
North Park Rd, Exeter
EX4 4QF, United Kingdom

We present a multimodal neural multi-net that models child language at the one-word and two-word stage. In this multi-net a modified counterpropagation network models one-word language acquisition whilst a temporal Hypermap models two-word language acquisition. The multi-net incorporates an exposure dependent probabilistic gating mechanism that predisposes it to output one word utterances during the early stages of language acquisition and to increasingly become predisposed to outputting two word utterances as the simulation progresses. The multi-net exhibits a gradual transition from the one-word stage to the two-word stage similar to that observed in children undergoing the same developmental phase.

1. Introduction

The Oxford English Dictionary defines the term *multimodal* as "characterised by several different modes of occurrence or activity; incorporating or utilising several different methods or systems". From this definition we can refer to *multimodal information* as information emanating from a single source that has been encoded into various modes. In this paper we present an unsupervised multimodal neural network model for early child language acquisition at the one-word and two-word stage informed by current thoughts in early child language development [1][2][3].

In contrast to earlier neural network models of child language acquisition that focus primarily on the one-word stage [4][5][6], the model we present in this paper is able to simulate the transition of early child language from the one-word stage to the two-word stage. Abidi and Ahmad [7] have previously presented a neural multi-net that is able to simulate both the one word and two word language acquisition stages. However, their model is not able to autonomously simulate the one-word to two-word transitional phase.

We have adopted the unsupervised learning paradigm for our model, in line with the current view that language acquisition in the natural setting is essentially an unsupervised self-organising process [3]. This is in contrast to early connectionist models of cognitive development such as the Plunkett, Sinha, Moller and Strandsby [4] model in which a neural network is trained through the backpropagation algorithm to simulate early lexical development. In addition, the model we present in this paper is informed by current opinion in neuroscience which suggests that information may be stored and processed in the brain using a common amodal representation [8][9][10]. This is in contrast to the predominant Hebbian-linked self-organising map [11][12] models of unsupervised cognitive processing that subscribe to the previously dominant view that multimodal information in the brain is primarily stored and processed by means of separate modality-specific modules that are linked to each other [13][14].

The rest of this paper is organized as follows: In the next section we give an overview of child language acquisition at the one-word and two-word stage. Then we discuss our counterpropagation network model for one-word child language acquisition. After this we present an overview of the temporal Hypermap, and show how it can be used for modelling language acquisition at the two word stage. Following this we present the gated multi-net model for simulating the transition from the one-word to the two word stage. Finally, we present a discussion on the results of the simulations and draw conclusions as well as suggesting ways in which the research on our language acquisition model can be taken forward.

2. Overview of Child Language at the One-Word and Two-Word Stages

Bloom [15] suggests that when an infant hears a word (and perhaps a larger speech unit like a phrase or sentence), the word is entered in memory along with other perceptual and personal data that include the persons, objects, actions and relationships encountered during the speech episode. In addition, it is now generally accepted by child language researchers that single word utterances at the one-word stage convey a child's communicative intentions regarding the persons, objects and events in the child's environment, and the conceptual relationships between them [1][2]. In our simulations, we model child language at the one-word utterance stage as tri-modal data comprising the actual one-word utterances, the perceptual entities and the conceptual relations that we in infer the child is expressing.

At the two-word stage children appear to determine the most important words in a sentence and, almost all of the time, use them in the same order as an adult would [16]. In addition, as can be deduced from Brown's set of basic semantic relations [17], it appears that children at the two-word stage use word utterances for pretty much the same reasons and under almost the same circumstances as infants at the one word stage. Consequently, as with our model of one-word child language, we model child language at the two-word stage as tri-modal data comprising the actual two-word sequences, the perceptual entities and the conceptual relations that we in infer the child is expressing.

3. Simulating One-Word Child Language

3.1. *The Modified Counterpropagation Network*

The full counterpropagation network [18][19] provides bidirectional mapping between two sets of input patterns. It consists of two layers, namely the hidden layer, trained using Kohonen's self-organising learning rule, and the output layer which is based on Grossberg's outstar rule. The Kohonen layer encodes the mapping between the two sets of patterns whilst the Grossberg layer associates each of the Kohonen layer neurons to a set of target output values. Each Kohonen neuron has two sets of weights, one for each of the two patterns being mapped to each other.

We use the Kohonen layer of the counterpropagation network to associate the corresponding input modal vectors. For a multimodal input comprising *m* modes, the Kohonen layer neurons will each have *m* modal weight vectors, with each vector corresponding to a modal input [Figure 1]. After training, when a modal input is applied to the network, the modal weights of the winning neuron will contain information on all the other modal inputs of the particular modal

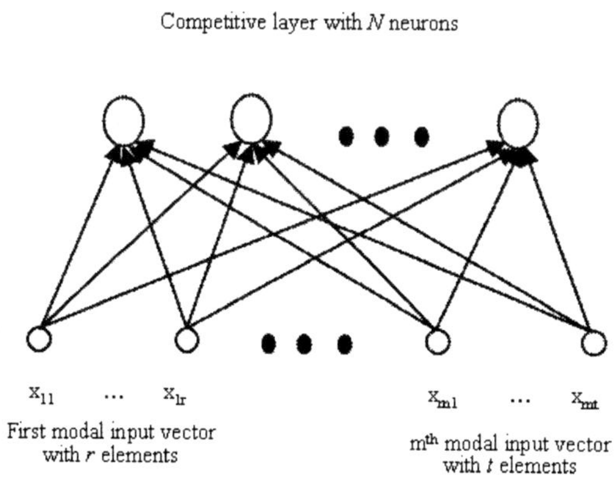

Figure 1: Multimodal Competitive Layer Adapted from the Full Counterpropagation Network.

input. By reading off these weights, we can get the corresponding modal inputs to a particular modal input.

For each multimodal input vector, the winning neuron is the one with the least overall Euclidean distance between its individual modal weight vectors and the corresponding modal component vectors of the multimodal input. To compute the overall Euclidean distance for each neuron we first determine the normalised squared Euclidean distance for each modal input:

$$d_j^2 = \frac{1}{n}\left\|\overline{x}_j - \overline{w}_j\right\|^2 = \frac{1}{n}\sum_{k=1}^{n}\left(x_{jk} - w_{jk}\right)^2 \tag{1}$$

where $\overline{x}_j$ and $\overline{w}_j$ are the modal input and weight vector respectively with n elements each. The overall Euclidean distance for the neuron is then obtained as follows:

$$D = \sqrt{\sum_{j=1}^{m} d_j^2} \tag{2}$$

Our counterpropagation network model of child language acquisition at the one-word stage actually encodes child utterances as composite multimodal elements comprising the phonological utterance, communicative intention and the referent perceptual entity. We believe that this approach simulates better the strong association between word and object, word and action, or word and event in early child language acquisition suggested by Bloom when compared to the Hebbian-linked self-organising map approach.

3.2. *One-Word Stage Model Simulation*

We trained a 10×10 modified counterpropagation network on the dataset over 500 cycles. The network node with the highest activation level was deemed to represent the response of the learnt association between the applied perceptual entity, conceptual relationship and utterance. The one-word utterance vector was then read from the winning network node, and the uttered word was determined from the training corpus using the nearest neighbour approach. In each situation, the network produced a similar response to the actual child's one word-utterance.

We used the training data every 10 epochs throughout the training period of 500 epochs to assess the network's ability to generate the correct word given a perceptual entity and a communicative intention as training progressed. As with the Plunkett, Sinha, Moller and Strandsby model [4], the network performance during training resembled children's vocabulary development during their second year. For instance, during the early stages of training, the network exhibited high error rates in generating the correct one-word utterances for input combinations of perceptual entity and communicative intention. However, as training progressed, the production of correct words suddenly increased until all the network was able to generate correct word for each of the situations presented to it.

Figure 2 shows a learning trajectory for a network with an initial learning rate of value 0.2. Increasing the learning rate caused the network to learn the 30 one-word utterances at a faster rate, and decreasing the learning rate also resulted in the network taking longer to muster the one-word utterances. However, for all values learning rate, the network still goes through an initial period of high error rate, followed by a period of lower error rate, which in turn is followed by a period of high error rate and finally by a period in which the error rate decreases continuously until the training set is mastered.

The learning trajectory for the network as training progresses suggests that children initially master some one-word utterances earlier on during learning, and then as the learning phase continues they undergo a period when the generation of correct one-word utterances deteriorates before the onset of the "vocabulary spurt" [4] where the error rate progressively decreases. The nature of the learning trajectory exhibited by our model is therefore consistent with the "U-shaped" developmental curves typical of child developmental activities such as language acquisition.

We also assessed the ability of our model to generalise to the correct one-word utterances following the input of a combination of perceptual entity and conceptual relation from a novel dataset of ten utterances independent of the

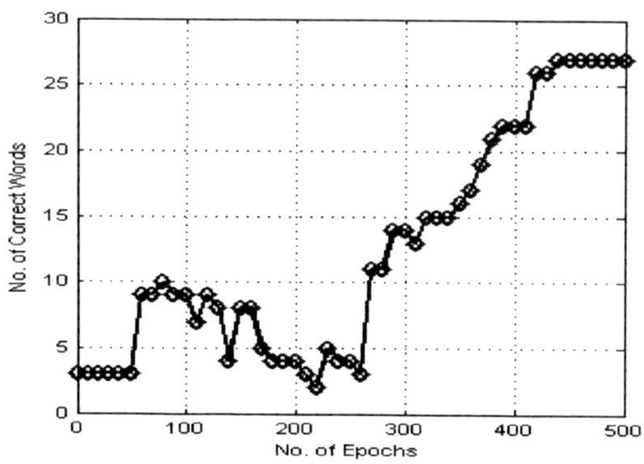

Figure 2: Plot of Correctly Recalled Words as A Function of the Number of Training Epochs.

training set. In nine of the ten cases, the modified counterpropagation network generalised to the correct one-word utterance. This suggests that the network successfully generalises to produce appropriate one-word utterances even for novel situations.

4. Simulating the Transition form One-Word to Two Word Language

4.1. *Temporal Hypermap Model for Two-word Simulation*

We simulate two-word child language acquisition using a temporal neural network based on the Hypermap architecture [20]. The temporal Hypermap [21] consists of a map whose neurons each have two sets of weights - context weights and pattern weights. When used for processing general sequences, the context weights identify the sequence whilst the pattern weights encode the sequence patterns. In our model, we use the pattern weights to encode the word utterances, and we subdivide the context weights to come up with a set of weights to encode perceptual entities and another set of weights to encode the conceptual relations that we infer the child is expressing. Associated with each neuron is a short term memory mechanism comprising of a tapped delay line and threshold logic units whose purpose is to encode the time varying context of the sequence component encoded by the neuron. This time varying context information makes it possible to recall an entire sequence based on a cuing subsequence. For instance, inputting the first word of an utterance will trigger the network to recall the entire word utterance sequence if the inputted word is unique to that sequence.

Consecutive neurons in a sequence are linked to each other through Hebbian weights, and inhibitory links extend from each neuron to all the other neurons coming after it in the same sequence. The Hebbian links preserve temporal order when spreading activation is used to recall a sequence, whilst the inhibitory links preserve temporal order when fixed context information is used to recall an entire sequence. When the multimodal vector constituting a stored sequence item is presented to the network, the Hebbian links enable all the other sequence items coming after the presented sequence item to be retrieved in their correct temporal order. For instance, when the perceptual entity and conceptual relation vectors are applied to the network along with the representation of the first word in an utterance, the remaining word utterance will be retrieved through spreading activation along the Hebbian links. And when only the perceptual entity vector and its associated conceptual relation vector are applied to the network, the entire word sequence is retrieved from the network in its correct temporal order by means of the inhibitory links.

4.2. *A Gated Multi-net Architecture for One-Word to Two-Word Transition*

We have created a gated multi-net system that comprises the counterpropagation network model for language at the one-word stage and the temporal Hypermap model for language at the two-word stage. During training, input vectors encoding the desired one-word utterance are presented to the modified counterpropagation network and at the same time the corresponding two-word sequence, along with its associated perceptual entity vector and conceptual relationship vector are presented to the temporal Hypermap.

At any one time, only one network is allowed to generate an output. Typically, the transition between child language acquisition stages is gradual and continuous [22]. We have modelled this transition with an exposure dependent probabilistic gating mechanism that predisposes the counterpropagation network to generate an output during the early stages of simulation, and the temporal Hypermap to generate an output during the latter stages of simulation. In this way, the model output changes gradually from one-word utterance simulation to two-word utterance simulation.

For each input, we make the likelihood of activating the temporal Hypermap in preference to the counterpropagation network a monotonically increasing function of the training cycle number. A simple function satisfying this requirement is the straight line equation:

$$y = mx + c \tag{3}$$

where y is the output, x is the input, m is the gradient, and c is the initial value of y.

For the gated multi-net, we replace the x term with the current training cycle number n, and we replace the constant c with the initial transition probability from one-word to two-word utterance prior to training, $p_u(0)$. The output y then gives the current transition probability $p_u(n)$ for the given input. At the end of training, the network responds with a two-word utterance to each input. Therefore the transition probability at the end of the training period is 1. If the total number of training cycles is n_T, the gradient m will be:

$$m = \frac{1 - p_u(0)}{n_T} \tag{4}$$

Hence, assuming a straight line relationship between the one-word to two-word transition probability and the number of the training cycle, the one-word to two-word transition probability is given by:

$$p_u(n) = (1 - p_u(0))\frac{n}{n_T} + p_u(0) \qquad (5)$$

4.3. *Gated Multinet Model Simulation*

From the Bloom 1973 corpus [1], we identify one-word and two-word utterances that seem to address the same communicative intention. However, a child's communicative intention at the one-word and two-word stage may not be exactly identical, as an analysis of the transcripts indicates. For a start, the transcripts seem to indicate that at the two-word stage the child has a greater ability to formulate and perceive relationships between more concepts in his or her environment than at the one-word stage. In addition, it appears that the child at the two-word stage interacts more with the objects in his or her environment than the child at the one-word stage. Thirdly, the transcripts indicate that the child's vocabulary at the two-word stage is larger than at the one-word stage, and the two-word utterances seem to indicate a deeper level of environmental awareness than can be ascribed to one-word utterances. Nevertheless, we have identified fifteen pairs of utterances that broadly match each other.

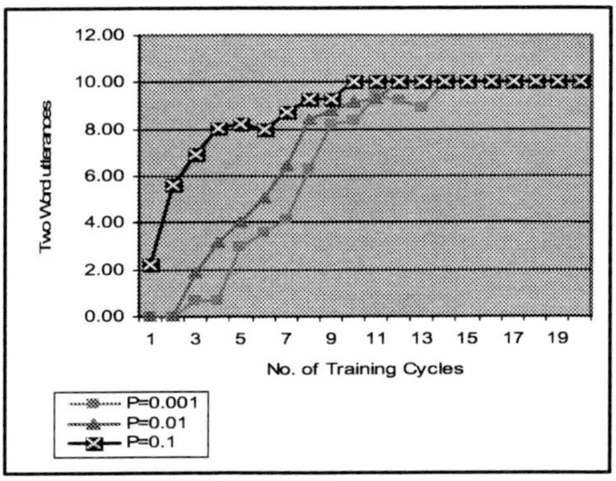

Figure 3: Output two-word utterances plotted against number of training cycles for the child language data set whose encoded events have the same frequency of occurence.

Our simulation of child language development from the one-word utterance stage to the two-word utterance stage shows that as the number of training cycles increases, the number of two word utterances increases proportionally, before reaching a saturation value independent of the initial transition probability (see Figure 3). The gated multi-net model therefore exhibits a gradual transition from one-word to two-word language utterance as seen in studies of child language acquisition.

In the gated multinet model of child language acquisition, the rate of increase of two-word utterances, prior to saturation, is dependent on the initial transition probability, with the higher the value of initial transition probability the larger the rate of increase of two-word utterances. Normal children also exhibit different rates of language development. Hence, by varying the initial transition probability value, we manage to simulate the variations in the rate of language development in normal children.

Our simulation of child language acquisition using the dataset with a frequency profile based on the limited environment to which Allison was exposed shows a steeper rate of increase of two-word utterances compared to the simulation using a dataset in which the events are equi-probable (see Figure 4). This suggests that the physically restricted environments to which infants are naturally exposed contribute towards quicker child language acquisition. Hence, our work lends support to Elman's suggestion that developmental restrictions on resources may constitute a necessary prerequisite for mastering certain complex domains [23].

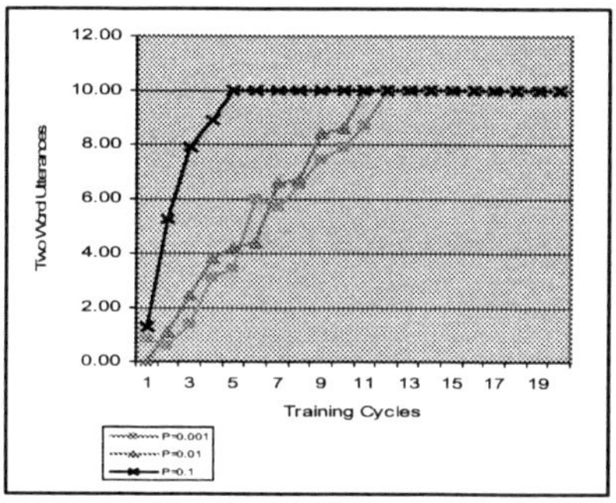

Figure 4: Output two-word utterances plotted against number of training cycles for the child language data set whose encoded events have different frequencies of occurence.

5. Conclusion and Future work

Our gated multi-net model of the transition of child language from the one-word utterance stage to the two-word utterance stage lends support to our conviction that unsupervised multimodal neural networks and unsupervised temporal neural networks provide a means for solving complex tasks that incorporate both multimodal and temporal characteristics. Like child language acquisition, most cognitive tasks can be viewed as both multimodal and temporal, and consequently, the field of cognitive modelling is likely to benefit from the approach we have adopted in out model.

However, whilst the gated multi-net model gives results that are consistent with child language data pertaining to the transition from the one-word stage to the two-word stage, it may be argued that such an approach is inappropriate since it gives the impression that these stages are implemented by different brain networks in the developing child. Rather, it would be more appropriate to assume that as the child's brain undergoes development, the networks implementing language processing progressively become better able to handle more complex language processing, hence the development from one-word language stage to the two-word language stage. This suggests that this developmental process is better modelled by a neural network architecture capable of adapting its structure to suit the structural and behavioural changes in the developmental data. As a consequence, we are currently investigating how we can use neural *constructivism* [24] to come up with a single unsupervised neural network architecture that can model the development of child language acquisition from the one-word to the two-word stage.

References

1. L. Bloom, One word at a time: The use of single-word utterances before syntax. The Hague: Mouton, 1973.
2. M. Small, Cognitive Development. San Diego: Harcourt Brace Jovanovich Publishers, 1990.
3. B. MacWhinney, "Models of the emergence of language," Annual Review of Psychology, vol. 49, 1998, pp. 199-227.
4. K Plunkett, C. Sinha, M.F. Moller, and O. Strandsby, "Symbol grounding or the emergence of symbols? Vocabulary growth in children and a connectionist net," Connection Science, Vol. 4, 1992, pp. 293-312.
5. P. Li, "Language acquisition in a self-organizing neural network model," in P. Quinlan, Ed., *Connectionist Models of Development: Developmental Processes in Real and Artificial Neural Networks*, Hove and New York: Psychology Press, 2003, pp. 115-149.
6. P. Li, I. Farkas, and B. MacWhinney, "Early lexical development in a self-organizing neural network," Neural Networks 17, 2004, pp. 1345 - 1362.
7. S.S.R Abidi and K. Ahmad Conglomerate neural network architectures: the way ahead for simulating early language development. Journal of Information Science and Engineering. Vol.13, 1997.: pp. 235 – 266.
8. A Caramazza, A. Hillis, B. Rapp, and C. Romani, "The multiple semantics hypothesis: Multiple confusions?" Cognitive Neuropsychology, vol. 7(3), 1990, pp. 161-189.
9. R. Vandenberghe, C. Price, R.Wise, O. Josephs, and R.S.J. Frackowiak, "Functional anatomy of a common semantic system for words and pictures," Nature, vol. 383, 1996, pp 254–256.

10. P. Bright, H. Moss, and L. K. Tyler, "Unitary vs multiple semantics: PET studies of word and picture processing," Brain and Language, vol. 89, 2004, pp. 417-432.
11. R. Miikkulainen, "A distributed feature map model of the lexicon," Proceedings of the 12th Annual conference of the Cognitive science Society, Hillsdale, NJ Lawrence Erlbaum, 1990 pp. 447-454.
12. R. Miikkulainen, "Dyslexic and category –specific aphasic impairments in a self-organising feature map model of the lexicon Brain and language, vol. 59, 1997, pp 334-366.
13. E.K.Warrington, "The selective impairment of semantic memory," Quarterly Journal of Experimental Psychology, vol. 27, 1975, pp 635- 657.
14. T. Shallice, "Specialisation within the semantic system," Cognitive Neuropsychology, vol.5, 1988, pp 133-142.
15. L. Bloom, The transition from infancy to language: Acquiring the power of expression. Cambridge University Press, 1993.
16. L.R. Gleitman. and Newport E. L. "The Invention of Language by Children: Environmental and Biological Influences on the Acquisition Language," In Gleitman, L.R. and Liberman, M. (Eds.). Language: An Invitation to Cognitive Science, Cambridge, MA: MIT Press, 1995. pp. 1-24.
17. R. Brown, A First Language: the early stages. London: George Allen and Unwin, 1973.
18. R. Hecht-Nielsen. Counterpropagation Networks. Applied Optics. Vol. 26,1987. pp. 4979-4984.
19. R. Hecht-Nielsen, "Counterpropagation Networks," Proceedings of IEEE International Conference on Neural Networks. Vol.2, 1987. pp.19-32.
20. T. Kohonen, "The Hypermap Architecture," In Kohonen T., Makisara, K., Simula O.and Kangas, J. (Eds). Artificial Neural Networks, Vol. II. Amsterdam, Netherlands, 1991. pp. 1357–1360
21. A. Nyamapfene, "Unsupervised multimodal neural networks," PhD. Dissertation, University of Surrey, Guildford, England, 2006.
22. J.H. Flavell, Stage-related properties of cognitive development. Cognitive Psychology, Vol 2, 1971. pp. 421 - 453.
23. J.L. Elman, "Learning and development in neural networks: The importance of starting small," Cognition, vol 48, no. 1, 1993, pp 71-99.
24. S. Quartz and Sejnowski, T. (1997). The neural basis of cognitive development: a constructivist manifesto. Behavioral and Brain Sciences Vol. 20 No. 4: pp 537-596.

CONSTRAINTS ON GENERALISATION IN A SELF-ORGANISING MODEL OF EARLY WORD LEARNING

JULIEN MAYOR* AND KIM PLUNKETT

Department of Experimental Psychology, University of Oxford, South Parks Road, Oxford, OX1 3UD, United Kingdom
** E-mail: julien.mayor@psy.ox.ac.uk*

We investigate from a modelling perspective how lexical structure can be grounded in the underlying speech and visual categories that infants have already acquired. We demonstrate that the formation of well-structured categories is an important prerequisite for successful generalisation of cross-modal associations such that even after a *single* presentation of a word-object pair, the model is able to generalise to other members of the category. This ability to generalise a label to objects of like kinds, commonly referred to as the *taxonomic assumption*, is an emergent property of the model and provides an explanatory framework for understanding aspects of infant word learning. Furthermore, we investigate the impact of constraints imposed on the Hebbian associations in the cross-modal training phase and identify the conditions under which generalisation does not take place.

1. Introduction

A central issue in early lexical development is how infants constrain the possible meanings of words to refer to objects of like kind. It is often assumed when infants learn a label for an object, they can apply it to the whole category. The generalisation of labels to new instances of objects within the same category is often referred to as the taxonomic assumption.[1] Many researchers have suggested that babies make use of this taxonomic assumption, along with a series of other constraints, in order to narrow the hypothesis space when learning new word-object associations.[2–4]

Markman proposed that even though infants find thematic relations between objects salient and interesting (e.g. dog and bone), a *taxonomic assumption* is used when generalising labels,[1] thereby overriding the thematic association. A series of studies have proposed that the taxonomic constraint is an evolved version of perceptually-based categorisation.[5,6] A notable example of which is known as the shape bias,[3,7] where infants show a clear preference to group items according to their overall shape.

Although there is a wealth of empirical data characterising these phenomena, very little is known about the nature of the neural mechanisms underlying them. We propose that such constraints are emergent properties of the underlying neural architecture. We use a model made out of two Self-Organising Maps connected together with associative links. Self-organising maps (often referred to as SOMs[8]) are good candidates for modelling the underlying mechanisms responsible for forming categories out of a complex input space; they achieve dimensionality reduction and auto-organisation around topological maps.[9]

Previous studies have highlighted the promising role of SOMs as models of early lexical development.[10,11] However, most of these studies have used heavily pre-processed input representations. In contrast we apply SOMs to *real* auditory and visual input using Hebbian learning to form cross-modal associations and examine in a computational framework how categorisation and generalisation can emerge. We will report results on the ability the network has to generalise word-object associations and discuss the implications for the taxonomic assumption.

We ran three experiments in order to investigate the constraints on the generalisation properties of the network. In experiment 1, we assess the role the number of word-object pairings plays in generalisation. We demonstrate that even following a single word-object pair presentation, the network is able to generalise the label to other images of like kind and vice-versa. Even though classification success increases along with the number of trained word-object pairs, the normalised generalisation is approximately independent of the number of pairings. This suggests that generalisation properties depend on the *physical* organisation of the model. Hence, we ran a second experiment in order to assess the influence of the map structure quality on generalisation capacity. We show that well structured maps are a prerequisite for good generalisation. Finally in experiment 3, we show how generalisation is limited by the number of units that are connected through Hebbian associations. Moreover, we show that generalisation performance reaches a peak even when only a limited number of units are allowed to *fire and wire* together, satisfying a constraint of limited synaptic resources.

2. Method

Our model consists of two SOMs each receiving input from one modality, either visual or acoustic. In a first phase of training, maps are *independently* fed with their respective input so that structure emerges. This first phase models the early experience of a baby discovering the environment, by sampling her visual and acoustic surroundings. In a second phase of training, we connect both maps with associative links. This second phase captures the increasing importance of shared attentional activities, such as gaze sharing, joint attention or pointing at objects, during later infancy. Although the integration of both maps probably occurs gradually in the real world, for the sake of simplicity we wait for the maps to be structured and then build associations. Through *simultaneous* presentation of both a visual token and an acoustic token that belong to the same category (e.g., mimicking the behaviour of a caregiver pointing at a dog while saying the word "dog"), synapses connecting active nodes on both maps are reinforced. After Hebbian associations between the two maps are formed, we test the model by presenting an image and measuring the activity patterns produced on the auditory map (or vice-versa). Successful generalisation occurs if unpaired images from the same visual category produce activation on the auditory map corresponding to tokens of the same label.

2.1. *Training the Unimodal Maps*

The algorithm of self-organisation is the standard Kohonen algorithm.[8] Each map (acoustic and visual) consists of an hexagonal grid of neurones receiving acoustic and visual inputs, respectively. With each neurone k is associated a vector m_k. For the presentation of each input pattern x, the vectors m_k are modified according to the following procedure:

We find the Best Matching Unit (BMU) i, defined by the condition

$$||m_i - x|| \leq ||m_j - x|| \qquad \forall j$$

By extension, we can identify the second best matching unit, the third, and so on. We apply a standard weight update rule with a learning rate that decays over time, $\alpha(t) = 0.05/(1 + t/2000)$ and a Gaussian neighbourhood function of the distance between neurones i and k, that shrinks in time $N(i,k)_t = e^{-||r_i - r_k||^2/2\sigma^2(t)}$. We define an averaged quantisation error, as a measure of weight alignment to the input, so that the Euclidian distance between input patterns and their respective best matching unit is:

$$E =< ||x - m_c(x)|| >_x$$

where $m_c(x)$ is the best matching unit for input pattern x. In order to shorten simulation time in experiments 1 and 3, we used a batch version of the algorithm.[8] In all experiments, map sizes were fixed to a 9x12 hexagonal grid of neurones for the visual map and to a 5x7 grid for the acoustic map.

2.2. *Coding the inputs*

2.2.1. *Image generation*

Images in the dataset are created from six pictures of animals (dog, cat, cow, fish, pig, sheep). Each image is first bitmapped into a square image having 20x20 pixels. We generate blurred versions of the six pictures in order to create multiple tokens in each category, centred on a prototype. We create 18 images per category by changing the grey scale value of a random number of pixels (min 0: max 400). The magnitude of the grey scale change is drawn from a normal distribution centred on zero and a standard deviation equal to 80% of the full grey scale. Prototypes are not included in the data set.

2.2.2. *On the importance of* real *acoustic token*

There is little consensus in the field as to what acoustic information babies use when identifying words. A series of studies emphasise the fact that babies pay attention to much more than simple features that would be described by a simple phonological encoding. In particular, it has been shown that at 9 months of age, babies are sensitive both to stress and phonetic information,[12] at 9.5 months they are able to make allophonic distinctions[13] and at 17 months, they pay attention to co-articulation.[14] All of these sensitivities to the speech signal may have an important impact on early lexical development. Therefore, we exploit the whole acoustic signature of tokens in order to avoid discarding relevant acoustic information.

We extract the acoustic signature from raw speech waveforms for six acoustic categories produced by nine female native speakers. By doing so, we confront the model with a lack of invariance in word pronunciation introduced by different speakers. Token are then normalised in length and sampled at regular intervals, 3 times per syllable[a]. After sampling, the sounds are filtered using the Mel Scale in order to approximate the human ear sensitivity. Input vectors are concatenations of three 7-dimensional mel-cepstrum vectors, derived from FFT-based log spectra[b].

---

[a]We found that for monosyllabic words, having 2 samples per syllable is sufficient from the point of view of word-object generalisation performance as described in the results section. We found no statistically significant improvement when increasing the number of time-slices beyond $N = 2$.

[b]The mel-cepstrum vectors are obtained by applying the following procedure: take the Fourier transform of a windowed excerpt of the signal, map the log amplitudes of the spectrum obtained above onto the Mel scale, using triangular overlapping windows and finally take the Discrete Cosine Transform of the list of Mel log-amplitudes.

2.3. *Forming the cross-modal associations*

After maps are structured following the presentations of the images and acoustic tokens in the data set, we mimic joint attentional activities between the care-giver and the baby by presenting simultaneously to both maps a randomly picked image from the data set and an acoustic token randomly picked within the matching category (e.g. one of the 18 images of dogs and one acoustic signature of a speaker saying the word "dog"). We build cross-modal associations by learning Hebbian connections between both maps. As a further simplification of the model, we use bidirectional synapses whose amplitudes are modulated by the activity of the connecting neurones. We define the neural activity of a neurone k to be $a_k = e^{-q_k/\tau}$ where q_k is the quantisation error associated with neurone k and $\tau = 5$ is a normalisation constant. There are several options for linking the maps:

- link *all neurones* on both maps
- link only *the* Best Matching Unit of the paired image on the visual map and *the* Best Matching Unit of the paired acoustic token on the acoustic map
- link together only a percentage of the neurones on both maps.

In experiments 1 and 2, only the top 25% of the Best Matching Units are linked together whereas in experiment 3 the percentage of units that are allowed to *fire and wire* is varied in order to investigate the role of this linking parameter when generalising word-object associations.

All synapses were first randomly initialised with a normal distribution centred on 1 and with a standard deviation of $\frac{1}{\sqrt{(1000)}}$. Synapse amplitudes are modulated according to a standard Hebb rule with saturation. Therefore synapse weights stay in a *physiological* range even for high neural activities. The synapse connecting neurone i from the visual map to the neurone j of the acoustic map is computed as follows:

$$w_{ij}(n+1) = w_{ij}(n) + 1 - e^{\lambda a_i a_j}$$

where n refers to the index of the word-object pairing and $\lambda = 10$ is the learning rate. The free parameters τ and λ were chosen by inspection to provide good results. After every word-object presentation, weights are normalised so as to model the limited synaptic resources: $\sum_{ij} w_{ij}^2 = 1$.

After training on cross-modal pairings we assess the capacity of the network to extend the association of a presented word-object pair to non-paired items that belong to the same category. Following a number of simultaneous presentations of word-object pairs, weights are fixed and all images in the dataset are classified according to whether the induced activity on the acoustic map corresponds to the activation of the appropriate label. This is referred to as the visual to acoustic condition: *v2a*. Averaging over all images in the data set gives us a measure which we call the *classification success*, C. We compare this measure to:

- the *perfect classification* C_{max}, achievable given the number of pairings (perfect classification in categories that "posses" a pairing, random classification in other categories)
- *item-based classification* C_{item}, where the items presented in pairs are associated perfectly, the other ones being classified at random
- the baseline where no learning occurs, the *random guess*, with one chance out of six to classify correctly the image.

We define a normalised value for *generalisation*, G, so that perfect classification given the pairings has a value of 1 and perfect memory with no generalisation, the *item-based condition*, would give a score of 0:

$$G = \frac{C - C_{item}}{C_{max} - C_{item}}$$

Similarly acoustic tokens are classified according to the activity induced onto the visual map, referred to as the acoustic to visual condition: *a2v*. All results reported are averaged over 65 independent simulations.

3. Results

3.1. *Generalisation as a function of number of pairings*

We report both classification success and its normalised version, the generalisation measure, as a function of the number of word-object pairs on which the network has been trained. We expect the network classification success to improve with an increasing number of label-object pairs. A positive correlation between classification success and the number of joint presentations of objects and their labels is shown in the left panel of Fig. 1. In both conditions, classifying images by their induced activity onto the acoustic map (v2a condition) and classifying labels by monitoring the induced neural activity onto the visual map (a2v condition), the network outperforms their respective "no generalisation" baselines. This indicates that even following the presentation of a single word-object pair, the network is capable of generalising the association to other objects and labels within the same category. The difference between conditions in overall classification success can be explained by the different levels of variance associated with the visual and the acoustic inputs.

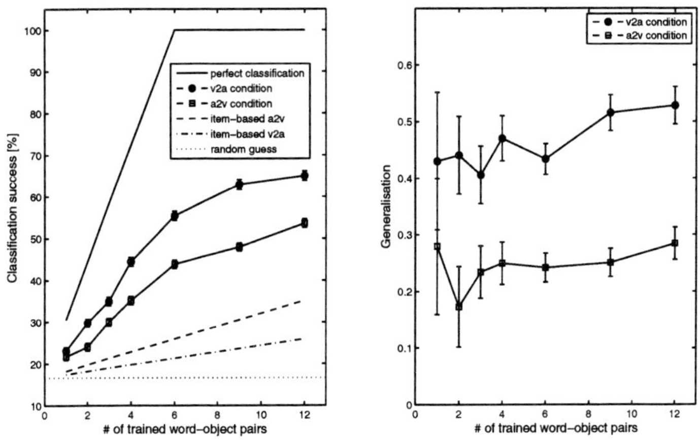

Fig. 1. Correct associations of labels to objects as a function of the number of simultaneous presentations of word-object pairs, after maps are structured. Left panel: Classification success for both conditions results are compared both to the maximal classification achievable (solid line) and to the results of a system only learning associations between paired tokens, with no generalisation capacity (dashed line for the a2v condition and dash-dotted line for the v2a condition). The dotted line corresponds to a random association of word and objects. Right panel: Generalisation as a function of the number of word-object pairings. Error bars correspond to one standard deviation after averaging over 65 simulations.

The right-hand panel of Fig. 1 depicts the normalised counterpart of classification success. We notice that to a first approximation there is no strong dependence upon the number of training pairs. In other words, when compared to both an optimal generalisation and to a simple item-based learning device, the network has a constant generalisation capacity. Even though the *absolute* ability of the network to associate objects with their labels increases along with the amount of joint word-object experience, the *relative* capacity of the network to generalise such associations is essentially independent of the number of such pairings. This indicates that the capacity of the network to generalise is dependent on the neural architecture, both at the level of the quality of the organisation of the maps and on the number of Hebbian associations that are allowed to *fire and wire* together. In the next experiment, we investigate the role played by the map qualities for the network's generalisation ability.

3.2. *Generalisation as a function of pre-pairing experience*

In order to investigate the role played by the neural architecture for generalisation capacity, we first controlled the quality of the maps structure before presenting the network with word-object pairs. Map structure improves with the experience. We monitored the average quantisation errors of both maps as a function of the number of times the whole data set is presented to the maps (defined as an epoch). In the bottom left panel of Fig. 2 we see the monotonic decline in quantisation errors for both maps as a function of increasing experience of images and sounds. In the top left panel of Fig. 2 we plot the classification success for the a2v condition as a function of pre-pairing experience, after training on 12 word-object pairs[c]. The top right-hand panel plots the same measure for the v2a condition.

In both panels, the classification success curves start from a random guess baseline to cross a level of performance comparable to that of an item-based learning network, before reaching a plateau in classification performance. When maps are still unstructured, neural activity is very low when an item is presented (the quantisation error is high). Hence, Hebbian learning is still too weak to be able to associate paired items reliably. When structure starts to emerge, presentation of word-object pairs elicit map activities sufficiently large to promote significant weight changes. At this stage in map development, object-label pairs are associated item-by-item but the lack of topological organisation in the maps is such that generalisation cannot be sustained yet. This stage of item-based performance corresponds to 100 epochs in the a2v condition and 35 epochs in the v2a condition. Finally, when the network has enough experience with items before pairings, maps are well organised and associations made during the pairing phase are generalised well to the other non-paired items.

In the bottom right panel of Fig. 2 we plot classification success as a function of the simple average of the quantisation errors of the maps. This comparison provides a more direct index of the impact of map structure on generalisation. The monotonic increase in generalisation quality as the maps structural quality increases (quantisation errors decrease) confirms

[c]Very similar results were obtained with 4 word-object pairs.

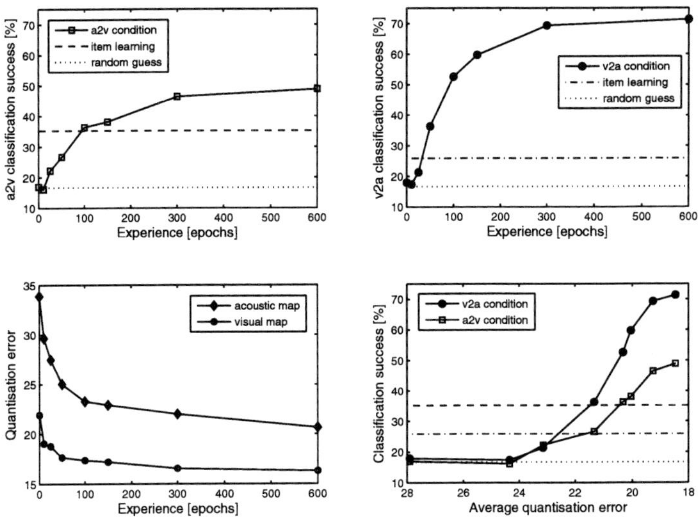

Fig. 2. Top row: classification success as a function of pre-pairing experience with objects and labels in the a2v condition (left) and v2a condition (right). Bottom left panel: quantisation error (map structure) as a function of experience. Bottom right panel: classification success as a function of quantisation error (map structure).

our claim that generalisation quality ultimately depends on the *pre-lexical* (pre-pairing) categorisation abilities.

3.3. *Generalisation capacity as a function of the number of Hebbian associations*

Finally we investigate the role played by the number of neurones allowed to be associated through Hebbian connections. Fig. 3 displays classification success as a function of the percentage of the neurones that are allowed to *fire and wire* together. When only one Best Matching Unit on each map is allowed to *fire and wire*, generalisation of labels to other images (and of images to other labels) fails. By increasing the number of nodes allowed to connect on both maps, we reach a maximal generalisation capacity. It is noteworthy that performance reaches a plateau when about 15 to 25% of the neurones are connected to each other. Additional capacity does not result in improved generalisation. The slow decay in the quality of generalisation past this maximum is explained by the penalty induced by introducing a greater number of weights.

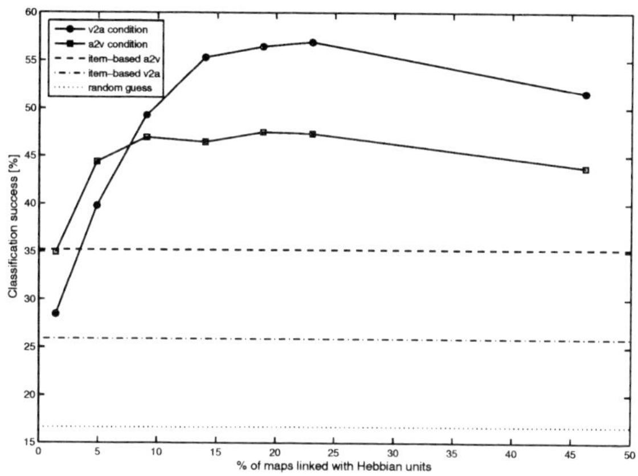

Fig. 3. Classification success as a function of the percentage of maps that are linked through Hebbian connections.

It might be argued that an autonomous procedure designed to identify the Best Matching Units is not biologically plausible. However, there are several potential solutions to this problem. First, the synapses that need to be reinforced are precisely those connecting neurones simultaneously with high activities. It is not unreasonable to suppose that a natural pruning procedure would eliminate the silent synapses so that only the strong ones would survive. This way only a limited fraction of nodes would be connected through associative links. Alternatively, because of the topological organisation of the SOMs, neurones representing similar items are close together. Hence, the BMUs are in the same region of the map and would not require a complex search procedure, satisfying both a limited synaptic resource argument and a locality rule.

4. Discussion

Connectionism is often considered to be an implementation of exemplar-based learning procedures. The way that supervised networks generalise novel inputs is through interpolation from the training set. Therefore, it is not guaranteed that such models will obey the taxonomic constraint. However, in our model, the first phase of training is completely unsupervised and we show that a single label-object pair can yield taxonomic responding (Exp. 1). The mechanism driving taxonomic responding is closely related to the percentage of neurones that are allowed to *fire and wire* (Exp. 3). If only one BMU from each map is allowed to *fire and wire*, taxonomic responding is not achieved. However, only a limited amount of synaptic resources is required in order to support good generalisation; we demonstrated in Exp. 3 that only about 20% of the maps need to *fire and wire* to achieve taxonomic responding.

The other prerequisite for good word-object generalisation in our model is to provide the system with well structured maps. In other words, successful word-object associations and their generalisations rely on pre-existing categorisation abilities. Once the perceptual system can achieve coherent object and sound categorisation, word-object associations are learnt fast and generalise well.

This finding is consistent with a series of studies suggesting that speech perception and cognitive development in infancy predicts language development in the second year of life.[16–19] Similarly, deficits in speech perception (bad auditory categorisation) predict language learning impairments[20] and more generally both delayed auditory perception[21] and impairments of visual imagery[22] are correlated with specific language impairments. The model captures the claim made that auditory perception bootstraps word learning.[23] We might also point out that the model predicts that visual categorisation bootstraps word learning in a similar fashion. Our findings suggests that generalisation of object-label associations are dependent upon good pre-lexical categorisation abilities and offers theoretical support for the experimental findings[16,17] that improvements in speech perception during infancy is an important developmental step toward language acquisition.

In summary, we show from a modelling perspective how taxonomic responding can be built on pre-existing categorisation abilities, along with limited synaptic resources. This neuro-computational account of taxonomic responding confirms the importance of pre-lexical categorisation abilities as predictors of successful lexical development.

References

1. E. Markman and J. Hutchinson, *Cognitive Psychology* **16**, 1 (1984).
2. E. M. Markman, *Categorization and naming in children: Problems of induction* (MIT Press, Cambridge, 1989).
3. B. Landau, L. B. Smith and S. Jones, *Cognitive Development* **3**, 299 (1988).
4. E. Markman, J. L. Wasow and M. B. Hansen, *Cognitive Psychology* **47**, 241 (2003).
5. D. Poulin-Dubois, I. Frank, S. Graham and A. Elkin, *British Journal of Developmental Psychology* **17**, p. 2136 (1999).

6. D. H. Rakison and G. E. Butterworth, *Developmental Psychology* **34**, 49 (1998).
7. S. Graham and D. Poulin-Dubois, *Journal of Child Language* **26**, 295 (1999).
8. T. Kohonen, *Self-organization and Associative Memory* (Springer, Berlin, 1984).
9. R. Durbin and G. Mitchinson, *Nature* **343**, 644 (1990).
10. R. Miikkulainen, *Brain and Language* **59**, 334 (1997).
11. P. Li, I. Farkas and B. MacWhinney, *Neural Networks* **17**, 1345 (2004).
12. P. Jusczyk, *Journal of Phonetics* **21**, 3 (1993).
13. P. W. Jusczyk, M. B. Goodman and A. Baumann, *Journal of Memory and Language* **40** (1999).
14. K. Plunkett, *Attention and Performance* **21** (In press).
15. M. Tomasello and J. Todd, *First Language* **4**, 197 (1983).
16. F. Tsao, H. Liu and P. Kuhl, *Child Development* **75**, p. 1067 1084 (2004).
17. P. Kuhl, B. Con boy, D. Padden, T. Nelson and J. Pruitt, *Language Learning and Development* **1**, p. 237 264 (2005).
18. J. MacNamara, *Psychological Review* **79**, 1 (1972).
19. R. F. Cromer, The development of language and cognition: The cognition hypothesis, in *New perspectives in child development*, ed. B. Foss (Penguin, Harmondsworth, 1974)
20. J. Ziegler, C. Pech-Georgel, F. George and C. Lorenzi, *PNAS* **102**, 14110 (2005).
21. L. Elliott, M. Hammer and M. Scholl, *Journal of Speech and Hearing Research* **32**, 112 (1989).
22. J. Johnston and S. Weismer, *Journal of Speech and Hearing Research* **26**, 397 (1983).
23. J. F. Werker and H. H. Yeung, *Trends in Cognitive Science* **9**, 519 (2005).

SELF-ORGANIZING WORD REPRESENTATIONS FOR FAST SENTENCE PROCESSING

STEFAN L. FRANK

Nijmegen Institute for Cognition and Information, Radboud University Nijmegen; and Institute for Logic, Language and Computation, University of Amsterdam Plantage Muidergracht 24, 1018 TV Amsterdam, The Netherlands E-mail: sfrank@science.uva.nl

Several psycholinguistic models represent words as vectors in a high-dimensional state space, such that distances between vectors encode the strengths of paradigmatic relations between the represented words. This chapter argues that such an organization develops because it facilitates fast sentence processing. A model is presented in which sentences, in the form of word-vector sequences, serve as input to a recurrent neural network that provides random dynamics. The word vectors are adjusted by a process of self-organization, aimed at reducing fluctuations in the dynamics. As it turns out, the resulting word vectors are organized paradigmatically.

Keywords: Word representation; Sentence processing; Self-organization; Recurrent neural network; Reservoir computing.

1. Introduction

There exist several psycholinguistic models that represent words as vectors in a high-dimensional state space. Distances between vectors encode strengths of relations between the corresponding words. Invariably, these are *paradigmatic* relations: Two vectors are close together in the state space if the represented words have a strong paradigmatic relation, that is, they belong to the same part-of-speech and/or have similar meaning. The best known models of this sort are Latent Semantic Analysis [1] and Hyperspace Analog to Language [2], but there are many others (for an overview, see [3]).

Such models have accounted for a considerable amount of experimental data regarding, among others, synonym judgement [1], lexical priming [4,5], vocabulary acquisition [1], and semantic effects on parsing [6]. This suggests that the mental lexicon is indeed organized paradigmatically, raising the question *why* this would be so. It seems unlikely that our mental lexicon has developed to make synonym judgement or lexical priming possible. Rather, it makes more sense to organize words in a manner that facilitates fast sentence processing and production. In such an organization, two word vectors would be close together if one words is likely to follow the other in a sentence. However, this constitutes a *syntagmatic* rather than a paradigmatic organization.

This chapter presents a connectionist model demonstrating that the two types of organization are in fact strongly related: A syntagmatic organization of word sequences is facilitated by a paradigmatic organization of individual words. That is, word representations encode paradigmatic relations because this allows for time-efficient sentence processing.

A related explanation for the nature of word representations was provided by the work of Elman [7], who trained a Simple Recurrent Network to predict which word would occur next at each point in a large number of sentences from an artificial language. During training, word representations were adapted to become more useful for this word-prediction task. As it turned out, the resulting organization was clearly paradigmatic. This suggests that a paradigmatic organization of words facilitates word prediction, which is presumably useful for sentence processing.

The model presented here is also a recurrent neural network that processes sentences from an artificial language, but differs from Elman's work in three respects. First, I assume that word representations are explicitly adapted to allow for faster sentence processing, and that word prediction only follows from this implicitly. In Elman's case, this relationship is reversed since his network was explicitly trained to perform word prediction, while it was left implicit how this is beneficial for sentence processing.

Second, word representations are adjusted by an unsupervised process of self-organization rather than by supervised backpropagation. In has been argued that, in the brain, unsupervised learning occurs in the cortex while supervised learning only takes place in the cerebellum [8]. Given ample evidence that word meanings are stored in the cortex, unsupervised learning is preferred for the current simulations.

Third, the weights of recurrent connections in the network are not adapted, making neural network training much more efficient. Current developments in recurrent network research have focused on so-called 'reservoir computing' [9–11], in which the recurrent part of the network is not trained but serves as a reservoir of complex dynamics that forms a task-independent memory trace of the input sequence. A non-recurrent network is then trained to transform the reservoir's activation states into target outputs. Only recently have such systems been applied to language processing [12–15]. The simulations presented here differ from other applications of reservoir computing in that learning is unsupervised, meaning that there are no target outputs and, therefore, no output connections to train. Instead, it is the input representations that are adapted.

The rest of this chapter is organized as follows: Section 2 describes the semi-natural language that was used in the simulations. Following this, Sec. 3 gives the details the model and the rationale behind the algorithm for adaptation of word representations. Simulation results are presented in Sec. 4 and discussed in Sec. 5.

2. The language

The artificial language used for the simulations was originally designed by Farkaš and Crocker [16] for training a network on the word-prediction task. All sentences of this language are also sentences of English but, of course, the language has much smaller vocabulary and simpler grammar.

2.1. *Lexicon*

There are 71 words in the language, as listed in Table 1. Note that the word 'who' serves both as a relative and as an interrogative pronoun. Additionally, there is a period symbol to mark the end of a sentence, making a total of 72 symbols.

2.2. *Sentences*

Words are combined into sentences according to a probabilistic context-free grammar (PCFG) that is too complex to be printed here in full. A simplification of the grammar is presented in Table 2. Note that noun-verb number agreement and some semantic constraints are not shown, but do apply nevertheless.

Sentences come in three types: declaratives, interrogatives, and imperatives. Declaratives can contain subject-relative clauses and object-relative clauses, which can also be nested. The average sentence length is 5.5 words, plus the period that ends each sentence.

Table 1. Words and word classes in the language.

Class	Subclass	Words
Noun	Proper	*John, Kate, Mary, Steve*
	Mass	*bread, meat, fish*
	Singular	*boy, cat, dog, girl, man, woman*
	Plural	*boys, cats, dogs, girls, men, women*
Verb	Singular	*barks, bites, chases, eats, feeds, hates, hears, likes, runs, sees, sings, swims, talks, walks*
	Plural	*bark, bite, chase, eat, feed, hate, hear, like, run, see, sing, swim, talk, walk*
	Auxiliary singular	*does, is ,was*
	Auxiliary plural	*do, are, were*
	Other	*wanna*
Adjective		*crazy, ferocious, good, happy, hungry, mangy, nice, pretty, sleazy, smart*
Article		*a, the*
Pronoun	Demonstrative	*that, those*
	Interrogative	*what, where, who*
	Relative	*who*

Table 2. Simplification of the PCFG for producing sentences. Items in square brackets are optional. N = singular or plural noun; N_{pr} = proper noun; N_{mass} = mass noun; V_{tr} = transitive verb; V_{in} = intransitive verb; Adj = adjective; Dem = demonstrative pronoun; Art = article.

Head		Production
S	→	Declarative . | Interrogative . | Imperative .
Declarative	→	NP [*who* RC] VP | NP V_{be} Adj | Dem V_{be} NP
Interrogative	→	Q_{wh}| Q_{aux}
Imperative	→	VP
NP	→	Art [Adj] N | [Adj] N_{pr} | N_{mass}
VP	→	V_{in} | V_{tr} NP
RC	→	V_{in} | V_{tr} NP [*who* RC] | NP [*who* RC] V_{tr}
Q_{wh}	→	*where/who* V_{be} NP | V_{do} NP V_{in} | *what* V_{do} NP *do*
Q_{aux}	→	V_{do} NP [*wanna*] VP | V_{be} NP Adj
V_{be}	→	*is* | *are* | *was* | *were*
V_{do}	→	*do* | *does*

3. The model

3.1. *The dynamical system*

We begin by taking the simplest possible discrete-time linear dynamical system:

$$\mathbf{x}_t = \mathbf{W}\mathbf{x}_{t-1} + \mathbf{y}_t,$$

where $\mathbf{x}_t \in \mathbb{R}^n$ is the system's n-dimensional state vector (with $\mathbf{x}_0 = \mathbf{1}$) and $\mathbf{y}_t \in \mathbb{R}^n$ the input at time step t. Matrix $\mathbf{W} \in \mathbb{R}^{n \times n}$ has values randomly chosen from a uniform distribution centered at 0, and is rescaled to have a spectral radius of 1. In neural networks terms, $\mathbf{x}_t$ would be the activation pattern over n units, $\mathbf{y}_t$ the input activation, and $\mathbf{W}$ the matrix of the recurrent network's connection weights.

A sequence of t input words (e.g., a sentence) corresponds to a sequence of t input vectors $\mathbf{y}_1, \ldots, \mathbf{y}_t$. The sequence $\mathbf{x}_0, \ldots, \mathbf{x}_t$ is the state-space trajectory resulting from that input. More specifically, each word w in the language's 72-symbol vocabulary is represented by a vector $\mathbf{v}_w \in \mathbb{R}^k$, with $k < n$. If word w is the input word at time step t, then the first k elements of input vector $\mathbf{y}_t$ equal $\mathbf{v}_w$, while the other elements are 0. In neural network terms, this means that only the first k of the recurrent network's n units receive input activation.

3.2. *Adjusting word vectors*

Initially, each word vector $\mathbf{v}_w$ has random values, uniformly distributed between ± 1. These vectors are adjusted on the basis of 5000 training sentences that were randomly generated from the PCFG of Table 2.

Let $w_1, \ldots, w_{t-1}$ be the training sentence so far (hence, $\mathbf{x}_{t-1}$ is the current state), and w_t the next word to occur. This provides some evidence that w_t is likely to occur after $w_1, \ldots, w_{t-1}$. In a syntagmatic organization of state space, this would mean that the state $\mathbf{x}_t$, resulting from input $\mathbf{y}_t$, is relatively close to $\mathbf{x}_{t-1}$ because, in such an organization, the nearness of two consecutive states mirrors the likelihood of the second state following the first. This consideration leads to the following informal rule for adjusting word vectors: Whenever $\mathbf{x}_{t-1}$ and $\mathbf{y}_t$ occur as a result of training input, input $\mathbf{y}_t$ is changed to $\mathbf{y}'_t$ such that the resulting $\mathbf{x}'_t$ is closer to $\mathbf{x}_{t-1}$ than $\mathbf{x}_t$ would have been. An even less formal way to put this is: Reduce the fluctuations of network activation resulting from the training input.

Formally, the learning rule is expressed by:

$$\Delta \mathbf{v}_w = \eta \left(\mathbf{x}_{t-1}^{(k)} - \mathbf{W}^{(k)} \mathbf{x}_{t-1}^{(k)} - \mathbf{v}_w \right), \tag{1}$$

where w is the word that occurs at time step t in the training sequence, $\mathbf{x}_{t-1}^{(k)}$ denotes the vector consisting of the first k elements of $\mathbf{x}_{t-1}$, $\mathbf{W}^{(k)}$ is the matrix consisting of the first k rows of $\mathbf{W}$, and $\eta = .001$ is a learning rate parameter.[a]

3.3. *Measuring syntagmaticity*

If the state-space trajectories indeed show a syntagmatic organization, trajectories resulting from grammatical sentence input should be shorter than those resulting from random word sequences. Syntagmaticity is therefore measured by comparing trajectory lengths resulting from grammatical sentences to those resulting from 'pseudo sentences'.

A set of 3352 test (i.e., non-training) sentences was fed through the system and the euclidean distances between all consecutive points in all resulting trajectories were summed to give the total trajectory length l_{test}. Next, pseudo sentences were constructed from the test sentences by randomly reordering the words, while leaving the end-of-sentence markers in place. This guarantees that pseudo sentences have the same length distribution and word frequencies as test sentences. Also, care was taken to make sure that word repetitions occur as often in the pseudo sentences as in test sentences. The extent to which the system is syntagmatic is now defined as

$$\text{syntagmaticity} = \frac{l_{\text{pseudo}}}{l_{\text{test}}},$$

where l_{pseudo} is the total trajectory length resulting from processing the pseudo sentences. Before training, there is no reason to expect any difference between l_{test} and l_{pseudo}, so the syntagmaticity level will be close to 1. If training is successful, syntagmaticity becomes larger than 1.

3.4. *Measuring paradigmaticity*

Word representations are organized paradigmatically if the vectors for words that belong to the same part-of-speech and/or have similar meaning, are closer together than vectors for words that are not paradigmatically related. The paradigmaticity of word vectors is measured by first defining classes of paradigmatically related words. There are 12 such classes, and they are exactly the 12 (sub)classes of Table 1 that contain more than one word.

[a]The reason why $k < n$ is that, otherwise, $\mathbf{x}_t = \mathbf{x}_{t-1}$ can easily be obtained for all t, by setting all $\mathbf{v}_w$ to $\mathbf{x}_0 - \mathbf{W}\mathbf{x}_0$. In other words, the recurrent units that do not receive input provide some 'noise' which cannot be compensated perfectly by adjusting the word representations.

Next, the average euclidian distances among vectors of words within a class (d_{within}) and between classes (d_{between}) are computed. The extent to which word vectors are organized paradigmatically is the ratio between the two:

$$\text{paradigmaticity} = \frac{d_{\text{between}}}{d_{\text{within}}}.$$

Initially, all word vectors are random so paradigmaticity will be close to 1. If word vector adjustment leads to a paradigmatic organization of the words, the measure for paradigmaticity will become larger than 1.

4. Results

4.1. *Parameter setting*

To investigate the effects of the dimensionalities of the state space and of the word vectors, the values of n and k were varied from 25 to 150, and from $.2n$ to $.9n$, respectively. There turned out to be no large qualitative effect of n. Syntagmaticity improved with larger k (albeit at the expense of paradigmaticity), which was to be expected since larger k means that more of the system's dynamics can be controlled by the adaptation algorithm that was designed to increase systematicity.

4.2. *Syntagmaticity and paradigmaticity*

Figure 1 shows how syntagmaticity and paradigmaticity develop during training, with parameters set to $n = 60$ and $k = 50$. As expected, syntagmaticity quickly rises above 1. This shows that the adaptation rule of Eq. (1) had the desired effect on syntagmaticity. After a few training cycles, however, syntagmaticity decreases slightly and levels of at around 1.28 (note the logarithmic scaling of the x-axis). More interestingly, the organization of word representations becomes strongly paradigmatic. Even after syntagmaticity has stabilized, the self-organizing process that was designed to increase syntagmaticity results in an increase in paradigmaticity instead. This is clear evidence for a link between the two types of organization.

4.3. *Word representations*

As Fig. 1 shows, the level of paradigmaticity more than doubles as a result of adapting word representations. It is not obvious, however, what this means in practice. Is the clustering of word vectors into meaningful groups

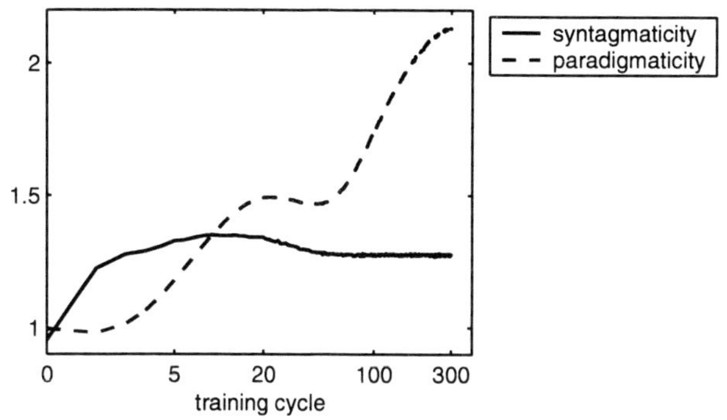

Fig. 1. The effect of training on syntagmaticity of state-space trajectories and paradigmaticity of word representations. In each training cycle, all 5000 training sentences are processed.

strong enough to be noticeable? In Figs. 2 and 3, the word vectors are plotted according to their first two principal components, which account for as much as 93.8% of variance. Signs of a paradigmatic organization are clearly visible. For instance, the four proper nouns cluster together, as do the singular verbs. However, there is also some evidence that the organization is incomplete, for example, the singular nouns do not seem to be clearly separated from the plurals. In retrospect, this is easy to explain: In Eq. (1), the change in $\mathbf{v}_w$ depends only on the previous inputs and not on what follows. Whether a singular or plural noun can appear, does not depend on the previous context, so the two subclasses will not be separated.

5. Discussion

The model's results clearly support the claim that words are represented according to their paradigmatic relations because this facilitates a syntagmatic organization of word sequences. The latter is useful for fast sentence processing and production, because it means that words that are likely to occur next are near the current position $\mathbf{x}_{t-1}$ in state space, so they can be accessed quickly.

Most likely, the paradigmaticity of word representation can be improved by changing Eq. (1) such that $\Delta\mathbf{v}_w$ comes to depend not only on the current word's previous context, but also on the following input word. Of course, it remains an empirical question whether the word representations constructed by such a model would explain more experimental data than a model like the current one, in which only the previous context is relevant.

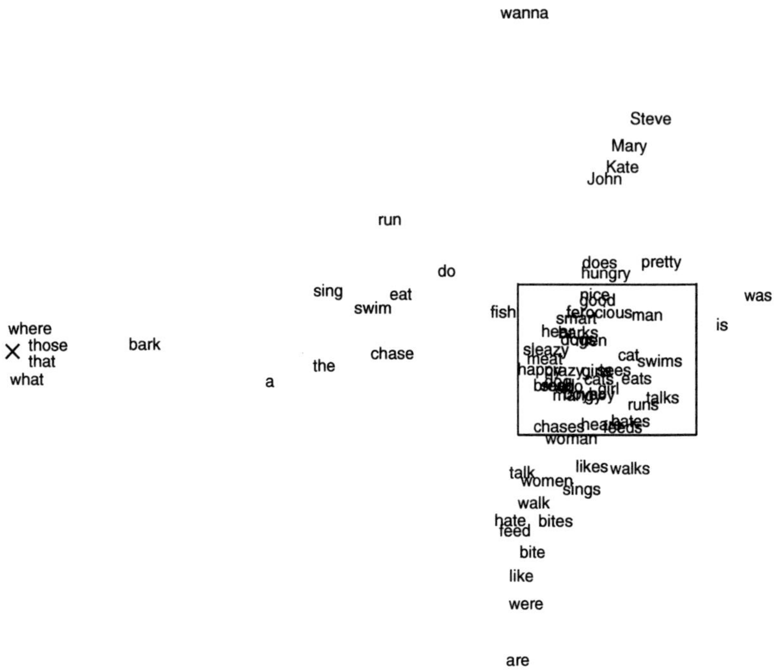

Fig. 2. Representations of the 71 words, projected onto the first two principal components. The '×' on the left-hand side indicates the position shared by the words *where*, *what*, *those*, and *that*. The area in the rectangle is shown enlarged in Fig. 3.

Fig. 3. Close-up of the rectangular area in Fig. 2.

If the distance between consecutive states $\mathbf{x}_{t-1}$ and $\mathbf{x}_t$ corresponds closely to the probability that the word that gave rise to $\mathbf{x}_t$ occurs in the context of $\mathbf{x}_{t-1}$, the model can be said to perform word prediction implicitly. Unlike Elman's network, the model does not give an explicit probability estimate for each word but such estimates could be derived from the organization of the state space.

If these word-probability estimates are accurate, it might be possible to use the model for predicting word-reading times. It has been argued by Hale[17] and Levy[18] that the time needed to read a word is proportional to its 'surprisal', which is simply the negative logarithm of its probability. If Hale and Levy are correct, and the model's state-space distances correlate positively with word surprisal, the model would predict word-reading times.

The sentences were generated by a known PCFG, so the probability of each word in each test sentence is available. However, the correlation coefficient between the negative logarithms of these probabilities and the model's state-space distances is only .27 (compared to .23 in advance of training), so the model cannot be said to be accurate enough to account for reading-time data. Considering that matrix $\mathbf{W}$ had fixed random values, such accurate predictions could hardly be expected. It is not unlikely that predictions could be improved by also adjusting $\mathbf{W}$ to the training inputs.

Acknowledgements

I would like to thank Igor Farkaš for kindly providing the sentences used in the simulations presented here, and Michal Čerňanský for useful discussions. This research was supported by grant 451-04-043 of the Netherlands Organization for Scientific Research (NWO).

References

1. T. K. Landauer and S. T. Dumais, *Psychological Review* **104**, 211 (1997).
2. C. Burgess, K. Livesay and K. Lund, *Discourse Processes* **25**, 211 (1998).
3. J. A. Bullinaria and J. P. Levy, *Behavior Research Methods* **39**, 510 (2007).
4. M. N. Jones and D. J. K. Mewhort, *Psychological Review* **114** (2007).
5. W. Lowe and S. McDonald, The direct route: mediated priming in semantic space, in *Proceedings of the 22nd annual conference of the Cognitive Science Society*, eds. L. R. Gleitman and A. K. Joshi (Mahwah, NJ: Erlbaum, 2000) pp. 806–811.
6. C. Burgess and K. Lund, *Language and Cognitive Processes* **12**, 177 (1997).
7. J. L. Elman, *Cognitive Science* **14**, 179 (1990).
8. K. Doya, *Neural Networks* **12**, 961 (1999).

9. H. Jaeger, Adaptive nonlinear system identification with echo state networks, in *Advances in neural information processing systems*, eds. S. Becker, S. Thrun and K. Obermayer (Cambridge, MA: MIT Press, 2003) pp. 593–600.
10. H. Jaeger and H. Haas, *Science* **304**, 78 (2004).
11. W. Maass, T. Natschläger and H. Markram, *Neural Computation* **14**, 2531 (2002).
12. S. L. Frank, *Connection Science* **18**, 287 (2006).
13. S. L. Frank, Strong systematicity in sentence processing by an Echo State Network, in *Proceedings of ICANN 2006*, eds. S. Kollias, A. Stafylopatis, W. Duch and E. Oja, LNCS, Vol. 4131 (Berlin: Springer, 2006) pp. 505–514.
14. S. L. Frank and W. F. G. Haselager, Robust semantic systematicity and distributed representations in a connectionist model of sentence comprehension, in *Proceedings of the 28th annual conference of the Cognitive Science Society*, eds. R. Sun and N. Miyake (Mahwah, NJ: Erlbaum, 2006) pp. 226–231.
15. M. H. Tong, A. D. Bickett, E. M. Christiansen and G. W. Cottrell, *Neural Networks* **20**, 424 (2007).
16. I. Farkaš and M. W. Crocker, Recurrent networks and natural language: exploiting self-organization, in *Proceedings of the 28th annual conference of the Cognitive Science Society*, eds. R. Sun and N. Miyake (Mahwah, NJ: Erlbaum, 2006) pp. 1275–1280.
17. J. Hale, A probabilistic Early parser as a psycholinguistic model, in *Proceedings of NAACL*, 2001 pp. 159–166.
18. R. Levy, *Cognition* (in press).

GRAIN SIZE EFFECTS IN READING: INSIGHTS FROM CONNECTIONIST MODELS OF IMPAIRED READING

GIOVANNI PAGLIUCA*

Department of Psychology, University of York, YO10 5DD York, UK

PADRAIC MONAGHAN

Department of Psychology, University of York, YO10 5DD York, UK

Spelling-sound correspondences in languages with alphabetic writing systems have two properties: they are systematic in that words that are written similarly have similar pronunciations, and they are compositional, in that letters or sets of letters within the written word correspond to certain phonemes in the pronunciation. The effects of systematicity and compositionality may vary for different psycholinguistic tasks and may also vary within the same task for different stimuli. We trained a neural network model to map English orthography onto phonology. Impairing the model in order to simulate a form of dyslexia (neglect dyslexia), revealed a distinction in performance within the naming task for words with different degrees of compositionality. The impaired model was sensitive to the presence of digraphs (multiletter graphemes that map into one phoneme, such as CH) in the word. The model shows sensitivity to the degree of compositionality, with words with digraphs less affected by damage than words without. We discuss these findings in light of a grain-size theory of reading.

1. Nature of the Reading Process

1.1. *Principles of the Mapping between Orthography and Phonology*

The reading process has generally been defined in the modeling literature in terms of forming a mapping between visual symbols and phonemes or syllables [4, 7, 26]. Learning to read can therefore be described as a process of learning to find shared "grain sizes" (or "what maps into what" pairs) between orthography and phonology in order to access the correct pronunciation of a word given its orthographic form [28].

* Corresponding author: gp517@york.ac.uk

The mapping between spelling and sound in languages with alphabetic writing systems has two general properties. It is *systematic*, in that words that are written similarly have similar pronunciations. It is also *compositional*, with individual letters, or pairs of letters, within the written word corresponding to certain phonemes in the pronunciation. These two principles of systematicity and componentiality govern the reading process across all alphabetic orthographies but are differently modulated by the nature of the mapping across different languages. In languages with shallow orthographies (such as Serbian or Italian) the phoneme associated with each letter can be processed almost completely independently of other letters in the word. In Italian for example, the letter A is always pronounce /α/ for every word in which it appears, regardless of the position in a word or the context (surrounding letters). The mapping between orthography and phonology for such languages can therefore be described as being highly systematic, with the *same* letter mapping (almost always) into the *same* sound, and highly componential, with *one* single letter mapping (almost always) into *one* single sound. In English instead, the pronunciation of the letter I varies according to the context: it is pronounced /ι/ in words like *mint* but /αι/ in words like *pint*. The relation between orthography and phonology is therefore less systematic in English than in Italian, with some letters or sets of letters being pronounced in different ways according to the context or the word they appear in.

The effects of systematicity and compositionality vary across languages. But they may also vary for different visual word processing tasks in the same language. Visual lexical decision, for example, is a task that requires the observer to decide whether a string of letter is familiar or not (whether it is a known word or not). In order to make the decision, the observer has to consider all the letters and map the whole orthographic form into a representation of the word that matches the given input. Single letters are therefore not independent but they are all necessary in order to perform the task successfully.

Semantic classification tasks also rely on different degrees of compositionality and systematicity to tasks involving naming. The meaning of a word is unrelated to its form (apart from a few well-attested exceptions of iconicity in language, see, e.g.[6, 16, 23]), the mapping between meaning and form being almost completely noncompositional and arbitrary. In order to access the meaning of a word, one needs to map more than the single letter into the semantic representation of the word, as expressed by Ziegler and Goswami, "[...] knowing that a word starts with the letters D tells [the child] nothing about its meaning" (p.1, [28]). A multi-letter mapping is applied to access the meaning of a word. Reading for pronunciation, then, entails a system that responds to the pronunciation of individual letters, or small groups of letters, within the word, whereas reading for meaning requires a system that processes the word in its entirety.

We have seen that the degree of systematicity and componentiality varies across languages and across tasks, but, furthermore, it may also vary *within* the same language for the same task. In English, the pronunciation of some letters can be determined solely from the letter alone, whereas for other letters pronunciation depends very much on the context. Multi-letter graphemes (e.g., th, sh) map into only one phoneme (/T/, /Σ/, respectively) instead of a purely compositional mapping forming two separate phonemes. In order to generate the correct pronunciation of the first phoneme in words like *thank* a larger shared grain size between orthography and phonology has to be considered that encompasses the single letter unit and takes into account a larger window of letter clusters over orthography, in this case the cluster TH that maps onto the phoneme /T/. A richer context must therefore be taken into consideration to generate the correct pronunciation when these clusters, or *digraphs*, are encountered.

In the following sections we introduce an original way of investigating these principles of systematicity and compositionality in reading, by means of investigating impaired reading performance observed in the neuropsychological syndrome of neglect dyslexia. We relate the neuropsychological data to a computational model of reading, and raise novel hypotheses about the labile nature of grain size in reading within a single language.

2. Neglect Dyslexia and the Reading System

Insights into the reading system and the principles that govern it can be derived from patients that demonstrate impairments to visual word processing. Neglect dyslexia is a reading impairment usually associated with right brain damage and unilateral visuospatial neglect. Patients with neglect dyslexia may fail to read verbal material on the left side of an open book, or the beginning words of a line of text, or more often the beginning letters of a single word [3, 9]. Neglect dyslexia has often been interpreted as an impairment of selective attention to the left visual field, but in the last few years a mounting body of evidence has suggested that partial information from the contralesional side is accessible at many levels of processing in these patients. Despite their very poor performance in reading words aloud, evidence for preserved processing of some aspect of lexical processing in the contralesional space has been shown by a series of investigations by Ladavas, Umilta' and Mapelli [14]. The authors found that Italian patients who could not read words or nonwords were nevertheless able to perform a correct lexical decision and semantic categorization (living-nonliving) judgments on the same stimuli. Neglect dyslexia therefore provides insight into the interaction and relative sparing of levels of lexical processing in the brain and the way different forms of mappings are affected by a single impairment.

In order to further explore the nature of the mapping between spelling and sound, we review the connectionist literature that dealt with this issue and present a simulation that aimed at investigating how this mapping is affected by damage to the system similar to the one observed in Neglect Dyslexia.

2.1. *Connectionist Models of Reading and Computational Characteristics of the Mapping*

An influential model of unimpaired reading is the triangle model [26]. The model was proposed as a challenge to the DRC approach of separate pathways for reading different types of words (a rule-based route for regularly pronounced words and a lexical route for irregular words, e.g.[4]). In place of specific representations of lexical items, the triangle model proposed instead that reading proceeded by interacting orthographic, phonological, and semantic representations of words, with all the representational modalities interacting with one another.

The triangle model was fully implemented by Harm and Seidenberg [8], where the model learned to map orthographic onto phonological and semantic representations through exposure to a large lexicon of English. Naming printed words was accomplished by the model through a process of "division of labor" between the direct orthographic-phonological route and the mediated orthographic-semantic-phonological route in the model. In addition, semantic processing of printed words, such as homophones, was mediated by the orthography-phonology-semantic route in the model. Thus, both orthography and semantics contributed to the fidelity of phonological representations in the model, and both contributions were required in order for processing to be completely accurate after training.

Critically, the orthography to phonology pathway and the orthography to semantics pathway are distinct in terms of the nature of the mapping between the representations: highly systematic for the orthography to phonology pathway, highly arbitrary for the orthography to semantics pathway. The computational implications of the different nature of arbitrary versus systematic mappings have been explored in terms of Age of Acquisition (AoA) effects in verbal tasks (better performance for early than late acquired items). Ellis and Lambon Ralph [5] showed AoA effects in a connectionist model trained with randomly-generated patterns that were abstractions of the properties of orthographic and phonological representations of words. The mapping between the input and output implemented was quasi-systematic, with a high, though not perfect, correlation between input and output patterns. The model exhibited different sensitivity to early versus late acquired items. Zevin and Seidenberg [27] noted that the relation between orthography and phonology is componential in nature, which was not respected in the Ellis and Lambon Ralph [5] model, and so Lambon Ralph and Ehsan [15] tested AoA effects in randomly-generated patterns that varied from being completely arbitrary (large AoA effects) to being entirely systematic and componential (no AoA effects). The nature of the mapping between input and output representations was used by Lambon Ralph and Ehsan [15] to predict differential behavioral outcomes for tasks that engaged arbitrary mappings between semantic and phonological representations compared to those that required translating orthographic onto phonological representations. Their model was found to predict AoA effects in picture naming (arbitrary mapping) but reduced effects in reading (componential and systematic mapping).

Impairment to visual attentional input associated with neglect results in damage principally to the leftmost portion of the word [18, 19]. Kinsbourne [11, 12, 13] proposed that each hemisphere of the brain attends to the visual field with a contralateral bias, such that there is a gradient of attention in each hemisphere, declining from left to right visual space for the right hemisphere and rising from left to right for the left hemisphere. In Kinsbourne's view, these gradients are equal and opposite and consequently attention is evenly distributed in the unimpaired system. When one hemisphere is damaged, however, attention becomes skewed contralesionally. Such a view has received support from investigations into the response properties of neurons in the parietal cortex of monkeys which have been found to follow a contralaterally-oriented gradient (for review see, e.g., [20]), and has been reinterpreted in terms of a neuronal gradient in the hemispheres [18, 21].

We hypothesized that a gradient of impairment to orthographic representations would result in a greater impact on naming than on lexical decision performance in neglect dyslexia patients, due to the computational properties of the mappings required for each task. Naming, as a systematic and compositional task, means that the pronunciation of the beginning of the word is somewhat independent of the pronunciation of the end of the word and so impairment of the first letters of the word is likely to have a profound influence on pronunciation of these first letters. In contrast, lexical decision, which may be interpreted as engagement of the orthographic to semantics mapping, requires information from all letters of the word to be integrated for generating the semantic representation. Consequently, impairment to the first letters of the word would have less of an impact on forming the semantic representation of the word as the mapping will be supported from letters elsewhere in the word.

Monaghan and Pagliuca [17] trained a version of the triangle model that learned mappings between orthography and phonology and semantics. After training, the model was impaired by degrading the orthographic input to the model along a gradient from left to right. The model demonstrated impaired reading of the left side of words, but the model′s lexical decision judgments, based on the words′ semantic representations, were relatively unimpaired, in a precise simulation of the patient data. The dissociation in the model resulted from the greater vulnerability to impairment of the regular, compositional mappings of the orthography to phonology pathway compared to the arbitrary, irregular mappings of the orthography to semantics pathway (words with very similar spelling can have very different meanings). Impairment to the mappings in computational models generally results in greater impairment to representations determined by less regular or more arbitrary mappings (e.g., for regular/irregular verbs: [10, 25]; for regular/irregular word naming: [7]). In our simulations of neglect dyslexia, the greater vulnerability of systematic mappings in the model was due to the precise type of impairment to the orthographic representations in the model, where there was a gradient of impairment from left to right over the orthographic units.

If highly compositional mappings appear to be more vulnerable to damage than arbitrary and less componential mappings following a graded lesioning of the input, then we should be able to observe a dissociation within a task if distinct grain sizes of processing are induced. Such a possibility is offered by the case of digraphs in the orthography to phonology mapping in English. We hypothesise that the grain size of processing for digraphs in English (e.g., CH, SH, TH) will be larger than for letter clusters that do not employ digraphs and are more compositional one-to-one mappings (e.g., CR, ST). Thus, we predict that words that contain digraphs (such as *chair* or *shovel*) should be more resistant to damage than words without digraphs (such as *cradle* or *stone*) in our model of neglect dyslexia.

2.2. *Modeling different Grain Size Processing in Neglect Dyslexia*

2.2.1. *Model architecture*

The network was comprised of an orthographic layer with 208 units, a hidden layer with 100 units, a phonological layer with 88 units and an attractor network (or set of clean-up units; see fig. 1). The orthographic units were fully connected to the hidden units, which in turn were fully connected to the phonological output units. All the connections were monodirectional. The orthographic representations for words were slot based, with one of 26 units active in each slot for each letter. There were 8 letter slots. Words were inputted to the model with the first vowel of the word in the fourth letter slot. So, the word "help" was inputted as "- - h e l p - -", where "-" indicates an empty slot. At the output layer, there were eight phoneme slots for each word. Three representing the onset, two for the nucleus, and three for the coda of each word, and so "help" was represented at the output as "- - h e - l p -". This kind of representation has the advantage of capturing the fact that different phonemes in different positions sometimes differ phonetically [7]. The phonological layer adapted a distributed representation of phonemes, with every unit corresponding to a phonetic feature, such as labial, sonorant or palatal. Each phoneme was represented in terms of 11 phonological features, as employed by Harm and Seidenberg [7].

The phonological attractor network was created by connecting all phonological feature units to each other and to a set of hidden units mediating the computation from the phoneme representation to itself (e.g., as in [7]). The direct connections between phonological units allowed the encoding of simple dependencies between phonetic features (the fact for example that a phoneme cannot be both *consonantal* and *sonorant*; e.g., [7]). The units were standard sigmoidal units with real-valued output ranging between 0 and 1 for the input (orthographic) and hidden layer, and –1 and 1 for the output (phonological) layer.

2.2.2. *Environment*

A set of 7291 monosyllabic word was used as training corpus. The words were selected from the CELEX English database [1]. Only words with frequency greater than 68 per million were in the database were selected. Each word was 1 to 8 letters long and was assigned log-transformed frequency according to its frequency in the CELEX database. Words with more than three phonemes in the coda were omitted from the input set.

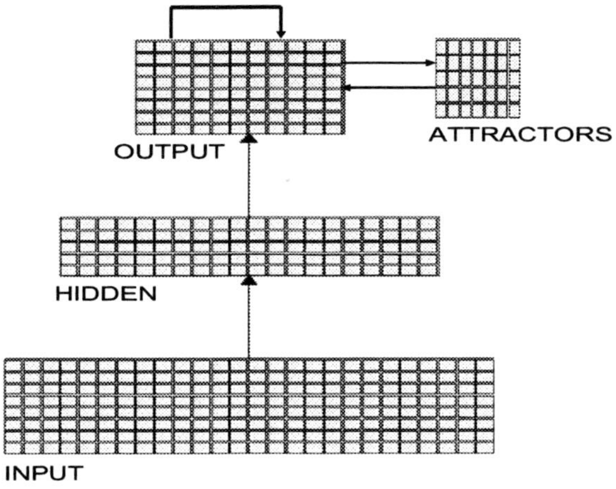

Figure 1. Architecture of the model. Input slots represent letters, output slots phonetic features.

2.2.3. *Training, Testing and Lesioning*

The model was trained using the backpropagation learning algorithm [24] the weight of the connections initialized to small random values (mean 0, variance 0.5) and learning rate μ=0.001. Words were selected randomly according to their frequency. Training was stopped after one million words had been presented.

For the reading task, the model's production for each phoneme slot at the output was compared to all the possible phonemes in the training set, and to the empty phoneme slot. For word presentations, if the model's performance matched that of the target phoneme representation for the presented word, then the model was judged to have read the word correctly. After training the weights over the connections were frozen and the model was tested and then lesioned.

In order to simulate damage to the right hemisphere of the brain, the activation from input letter slots was reduced along a gradient from left to right, such that the largest reduction in activation was from the leftmost letter slots. Two severities of lesioning were used, in order to simulate a mild impairment and a severe impairment.

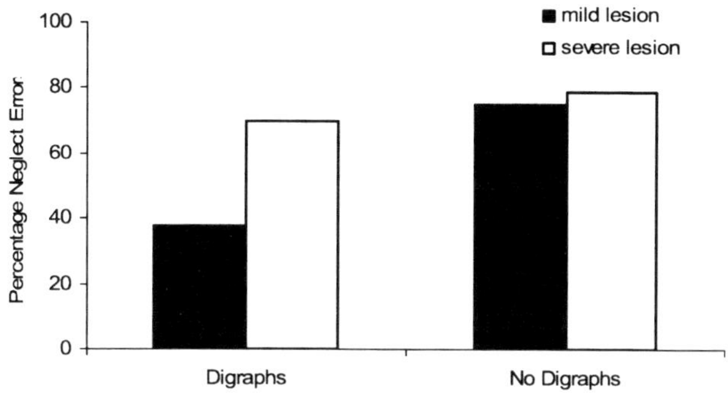

Figure 2. Proportion of neglect error for words with and without digraphs as a function of lesion.

2.2.4. *Results*

After 10 million patterns had been presented the model correctly reproduced 93.6% of the words in the corpus, which was a level comparable to the accuracy achieved by Harm and Seidenberg's [7] model of reading.

The model was then lesioned and tested over all the stimuli again. When the lesioned model made a reading error, it was usually an omission or substitution of the first phoneme. The severe model made errors on reading 83.5% of the words. Of these, 75% of the word-reading errors were classified as neglect errors. The mild lesion model made errors on 56.3% of the words. Neglect errors accounted for 66.2% of the errors. Both models were tested on a subset of words that presented digraphs in initial beginning (*shame*) and on a subset of control words without digraphs (*stain*). Three digraphs in initial position were considered: *ch*, *sh*, and *th*. The words were 4 to 7 letters long, matched for length, first letter, first and second bigram frequency, word frequency and were scored in the same way as the other words read by the model (correct responses, neglect errors, non neglect errors).

Results are shown in figure 2. Both the severe lesion and the mild lesion models show sensitivity to the presence of digraphs in initial position of word, with words with digraphs being less affected by the damage over orthography than words without digraphs. Overall words with digraphs present less neglect errors than words without digraphs (for the mild lesion, χ^2 (1) = 95.73, $p < .001$; for the severe lesion, χ^2 (1) = 6.67, $p < .05$). This result is in line with our prediction that the nature of the mapping (more or less componential) plays a role in the processing of the stimuli and affects performance under noisy conditions. The model indicates that grain size processing varies according to the stimulus characteristics, and that this is revealed through impairment of the model. Most importantly the model generates an empirically testable prediction: neglect dyslexia patients might show the same trend and produce less neglect errors in response to words with digraphs than to words without digraphs.

3. Conclusion

The nature of the mapping between orthography and phonology has been at the center of the present investigation. This paper aimed to explore the effects of different degrees of compositionality on the performance of a connectionist neural network trained to read English words and lesioned in order to simulate neglect dyslexia. We showed that highly compositional mappings are more vulnerable to damage than less compositional mappings, as observed when the lesioned model was tested on words with digraphs and without digraphs. Previous modeling work [17] suggested that different language tasks might imply different degrees of systematicity and compositionality that are differently affected by damage and can potentially account for dissociations observed in the performance of neglect dyslexic patients on the reading task and the lexical decision task. We have shown here that these principles apply not only between tasks, but also within the same task (reading) and that connectionist models of impaired reading can demonstrate sensitivity to these properties and guide empirical research.

Acknowledgments

This work was supported by an EU Sixth Framework Marie Curie Research Training Network Program in Language and Brain: http://www.ynic.york.ac.uk/rtn-lab

References

1. Baayen, R.H., Piepenbrock, R., & Gulikers, L., *The CELEX Lexical Database* (Release 2) [CD-ROM]. Philadelphia, PA: Linguistic Data Consortium, University of Pennsylvania (1995).
2. Behrmann, M., in M.J. Farah & G. Ratcliff (Eds.), *The neuropsychology of high-level vision.* Hillsdale, NJ: Lawrence Erlbaum Associates (1994).
3. Bisiach E., & Vallar G., Unilateral neglect in humans. In F. Boller, J. Grafman, & G. Rizzolatti (Eds.), *Handbook of Neuropsychology*, Amsterdam: Elsevier Science, 459 (2000).
4. Coltheart, M., Rastle, K., Perry, C., Langdon, R., & Ziegler, J., *Psychological Review*, **108**, 204 (2001).
5. Ellis, A.W. & Lambon Ralph, M.A., *Journal of Experimental Psychology: Learning, Memory & Cognition*, **26**, 1103 (2000).

6. Gasser, M., Sethuraman, N., & Hockema, S., in S. Rice and J. Newman (Eds.), *Experimental and empirical methods*. Stanford, CA: CSLI Publications (2005).
7. Harm, M.W., & Seidenberg, M.S., *Psychological Review*, **106**, 491 (1999).
8. Harm, M.W., & Seidenberg, M.S., *Psychological Review*, **111**, 662 (2004).
9. Hillis, A.E., & Caramazza, A., *Neurocase*, **1**, 189 (1995).
10. Joanisse, M.F. & Seidenberg, M.S., *Proceedings of the National Academy of Sciences*, **96**, 7592 (1999).
11. Kinsbourne, M., *Transactions of the American Neurological Association*, **95**, 143 (1970).
12. Kinsbourne, M., *Advances in Neurology*, **18**, 41 (1977).
13. Kinsbourne, M., in I.H. Robertson & J.C. Marshall (Eds), *Unilateral neglect: Clinical and experimental studies*. Hove, UK: Lawrence Erlbaum (1993).
14. Làdavas, E., Umiltà, C., & Mapelli, D., *Neuropsychologia*, **35**, 1075 (1997).
15. Lambon Ralph, M. & Ehsan, S., *Visual Cognition*, **13**, 928 (2006).
16. Monaghan, P. & Christiansen, M.H., *Proceedings of the 28th Annual Conference of the Cognitive Science Society*. Mahwah, NJ: Lawrence Erlbaum Associates (2006).
17. Monaghan, P., & Pagliuca, G. (submitted).
18. Monaghan, P. & Shillcock, R., *Psychological Review*, **111**, 283 (2004).
19. Mozer, M. C. & Behrmann, M., *Journal of Cognitive Neuroscience*, **2**, 96 (1990).
20. Plaut, D.C., McClelland, J.L., Seidenberg, M.S., & Patterson, K.E., *Psychological Review*, **103**, 56 (1996).
21. Pouget, A. & Driver, J., *Current Opinion in Neurobiology*, **10**, 242 (2000).
22. Pouget, A. & Sejnowski, T. J., *Psychological Review*, **108**, 653 (2001).
23. de Saussure, F., *Cours de linguistique générale*. Paris: Payot (1916).
24. Rumelhart, D.E., Hinton, G.E., & Williams, R.J., *Nature*, **323**, 533 (1986).
25. Rumelhart, D.E. and McClelland, J.L., in: MacWhinney, B., Editor, Mechanisms of language acquisition, Erlbaum, Hillsdale, NJ (1987).
26. Seidenberg, M.S., & McClelland, J.L., *Psychological Review*, **96**, 523 (1989).
27. Zevin, J. D. & Seidenberg, M. S., *Journal of Memory and Language*, **47**, 1 (2002).
28. Ziegler, J., & Goswami, U., *Psychological Review*, **131**, 3 (2005).

USING DISTRIBUTIONAL METHODS TO EXPLORE THE SYSTEMATICITY BETWEEN FORM AND MEANING IN BRITISH SIGN LANGUAGE

JOSEPH P. LEVY AND NEIL THOMPSON

School of Human & Life Sciences, Roehampton University, UK

We describe methods for measuring the correlation between gestural form similarity and meaning similarity between pairs of words/signs in the established lexicon of British Sign Language. Using a notation from a sign language dictionary we develop a tree-based representation that is capable of distinguishing the signs in the dictionary. We explore several different similarity/distance measures for the gestural forms and, using semantic representations developed elsewhere, we calculate the correlation between the respective form and meaning distances for all the combinations of pairs of words. Our results demonstrate that BSL exhibits a small but significant correlation between form and meaning similarities comparable to results found for spoken languages.

1. Introduction

It is widely accepted that, on the whole, there is no systematic relationship between the form of a word and its meaning. This has become known as Saussure's doctrine of the arbitrariness of the form [1]. It is seen as a hallmark property of natural language. Using Plaut & Shallice's [2] examples: the form (spoken phonology or written orthography) of the word "cat" is close to "cot" but they do not overlap in terms of their semantics. Conversely, "cat" is close to "dog" semantically but not in terms of any shared features of form. This arbitrariness in the way that form is mapped to meaning allows a great deal of flexibility for the coining and storage of new words but means that knowledge of other words or of perceptual features of the world cannot be used to help identify a word or infer the meaning of a new word.

There are several examples of exceptions to this principal of arbitrariness, including onomatopoeia, morphological affixes, expressives [3] and classifiers (markings for aspects of meaning such as form or function as seen in languages such as Japanese and BSL).

Although there is now broad agreement that sign languages are fully fledged forms of natural language, there is a widespread intuition that they are more "iconic" than spoken languages. In order to stress that sign languages are not in any way inferior to spoken languages, sign-language linguistics may have underemphasised the degree of systematic relationship between the form and meaning of individual signs. Taub [4] also suggests that the degree of iconicity in spoken languages may have been underestimated. In this chapter, we develop a distributional method of estimating one form of systematic relationship between form and meaning in British Sign Language (BSL). In order to measure a quantitative measure of systematicity across a large sample of lexical items, we examine the relationship between form-form and meaning-meaning similarities in pairs of signs.

Contrary to popular belief, there is no universal sign language. An example that helps to illustrate this fact is that British Sign Language, American Sign Language (ASL) and Irish Sign Language, despite being used in Deaf Communities within societies where spoken English is a dominant form of communication, are mutually incomprehensible by their respective users [5]. Sign languages are used by Deaf communities as native natural languages. They use complex patterns of hand shape and movement as well as non-manual features such as facial expression in an analogous way to spoken phonology to communicate lexical and sentential meanings. They are clearly not simple mime systems but linguistic systems that use conventions, patterns and apparent form-meaning arbitrariness. Spoken and signed languages are strikingly similar in aspects of development and the neurology underlying acquired deficits [6].

Distributional methods, where use is made of structural patterns of form and patterns of word usage that reflect aspects of meaning, provide a potential method for calculating a quantitative estimate of the relationship between form-similarity and meaning-similarity. Shillcock *et al* [7, 8] describe a method for giving a quantitative measure for such systematicity in spoken English. They encoded the phonological form of 1733 monosyllabic monomorphemic words of spoken English using eight distinct features used in a text-to-speech system [9]. Thus, their representation of form is based solely on a representation designed to capture the distinctions between the sounds of English words. They represented lexical semantics using high-dimensional vectors derived from the co-occurrence of written and spoken words in a large corpus of UK English [10]. They then calculated the pairwise similarities between the forms and meanings of every pair of words ((1733 x 1732)/2 = 1,500,778 word pairs). The phonological similarities were calculated as edit distances where the number of steps required to change one phonological form to another were counted with a set of weightings to adjust for the judged computational expense of different operations. Semantic "distance" was calculated as 1 – vector cosine between the word vectors, a method that has been found to correlate with other measurements and judgments of semantic similarity. [11]. They then simply calculated the Pearson product-moment correlation coefficient between the form- and meaning measures of the word pairs obtaining a value of $r = 0.061$. This small correlation was shown to be statistically significant using Monte Carlo (random permutation) techniques to demonstrate that the frequency with which form and meaning distances between random pairings of words produced values this high was small ($p < 0.001$).

Using similar techniques, Tamariz [12, 13] demonstrated comparable results for a small corpus of spoken Spanish. The main methodological difference was the use of the information theoretic Fisher distance as a measure of correlation and the Mantel method of shuffling for the Monte Carlo statistics.

Although these measured correlations are very small, the authors argue that their statistical significance is due to traces of systematicity in the lexicon between form and meaning similarities. Shillcock et al. [8] demonstrate that the correlation is not purely generated by function words or high frequency words and that it increases for polysyllabic words. Tamariz argues that the measured systematicity reflects adaptive pressures on the structural lexicon such as ease of learning and distinctiveness.

Taub [4] has argued that, far from making sign languages any less expressively powerful, the extra degrees of freedom afforded by the use of the temporal and spatial gestures of sign languages allow rich metaphorically mediated mechanisms for the coining of new lexical items.

Gasser [3] usefully distinguishes *absolute* iconicity from *relative* iconicity. The former is a direct relationship between word form and the world. In terms of mental computation, such a relationship may aid the coining of a new word or the inference of the meaning of an unknown word/sign by linking its form to knowledge about the world. Another function of systematicity between form and meaning (non-arbitrariness) would be to allow the inference of the meaning of an unknown word/sign from its similarity to a known word/sign. This language-internal kind of systematicity is an example of Gasser's relative iconicity. However, this kind of systematicity need not be what is generally understood as *iconic* since it need not transparently map onto any aspect of the world. For example if the concept of a hammer was expressed by the word "frod", the concept of mallet could be expressed by "crod" in a way that could be described as relative form-meaning systematicity without any absolute iconicity linking form to the world.

It is difficult to imagine a method for measuring absolute iconicity across a large number of words/signs. However, the method explored by Shillcock *et al* [7] and Tamariz [13] does provide a way of estimating relative form-meaning systematicity and we adapted it to estimate the degree to which British Sign Language exhibited this property.

2. Encoding BSL sign forms

We used 1545 signs given in the *Faber Dictionary of British Sign Language* [14]. These were all the signs that were coded and that also had a corresponding entry in the semantic vector database.The dictionary represented the gestural form of each sign as a symbolic notation that can be expressed in tree-form (see Figure 1).

The transcription of each sign's gestural form employed a system related to those that have been used for other sign languages, notably inspired by Stokoe's [15] work on ASL. A sign is produced using one or two hands. Each hand uses one of 57 conventional handshapes (*dez*) classified into 22 broad handshape families which are sub-divided using 9 diacritics to denote departures from the main handshape such as a protruding thumb. The positions of the handshapes (*tab*) are transcribed by another parameter (36 options) and denote where the hand is held (e.g., under the chin or on the left side of the chest). There are then also 6 possible palm orientations and 6 possible finger orientations. Hand movement (*sig*) is encoded as one of 30 broad movement categories and 5 modifiers. A modifier might, for example, distinguish a twist at the wrist in terms of how sharp the movement was. Movements can be simultaneous when two hands are being used and can involve a change in handshape. Movements can also be layered on top of each other, e.g., an index finger making a circular movement could at the same time also be moving forward. Signs can consist of a single handshape, two handshapes where the non-dominant one is stationary or two moving handshapes. There are 9 possible hand arrangements defining the positional arrangements of the hands when two are used. In addition to the transcription of manual gesture, the dictionary also classifies 46 non-manual features such as facial expression. These may also be in sequence or occur simultaneously. This formalism is used to describe the citation forms of the 1739 *frozen* or *established* signs of the dictionary. BSL also includes devices to generate a *productive* lexicon where gestural features act as linguistic classifiers to denote properties such as object shape or function. These open-ended lexical items would increase any quantitative estimate of the systematicity between form and meaning similarity but we restrict our initial study to the established lexical items in the dictionary. A quantitative analysis of everyday BSL signs would require an extensive transcribed corpus.

The psychological validity of this kind of gestural form feature system is discussed below where we describe how form similarity is computed. In order to cover the complete BSL dictionary, we encoded the above feature representation as a tree (see Figure 1). A complex sign might consist of more than one simpler sign. Each simple sign consists of at least one hand gesture and may also include non-dominant hand gestures and non-manual features. Each hand gesture is encoded as a starting hand-shape and position and a number of movements of that hand. Hand movements may be in sequence or simultaneous. If two hands are used, their mutual arrangement is encoded. Lastly, non-manual features may also be simultaneous with each other and the rest of the sign. Figure 1 shows how the above classification is described by the branches and leaves (features) of a tree where "sim" denotes simultaneous movements of a gesture and "mod" some further optional classification features for that particular level of the tree. All branches terminate in feature values but some of these have been omitted in the diagram for clarity. A bracketed node and its children are optional.

The tree structure captured the linguistic intuitions of the dictionary compilers in a way that appeared unbiased with respect to how form-similarity might correlate with meaning similarity.

This structure for BSL gestural form is one that has been developed using the same linguistic principles used in spoken language phonology. Atomic units of form are proposed if they distinguish between words/signs. This is seen in minimal pairs of phonemes that distinguish between different spoken words due to differences in a small number of features, e.g., the spoken words "cat" and "cot" are distinguished by the small differences between the two vowels such as whether the front or back of the tongue is used and the BSL signs for "bird" and "talk" are distinguished by handshape alone [5]. Computational studies of spoken word phonology generally use simplified phonetic feature systems that capture the main dimensions that distinguish words of a particular language. The Festival system [9] employed by Shillcock *et al.* consists of eight features. Since this is based on a speech synthesis technology it is easy to demonstrate that the feature-based representation gives a good approximation to an encoding of natural speech. The gestures of a sign language like BSL are more complex because there are more dimensions of articulation (e.g., two hands plus non-manual features, wider degrees of freedom for movement and placement of articulators, and more opportunity for simultaneity of gestures). A good approximation for the form of single spoken words can be encoded by a relatively simple feature vector. Although, the encoding for single signs is

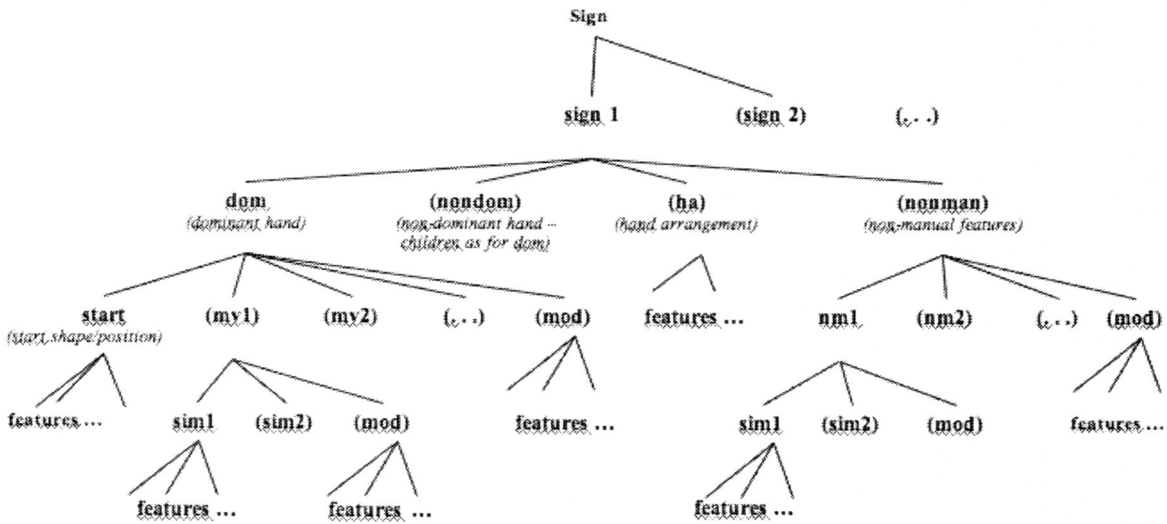

Figure 1: A tree structure representation of the gestural encoding used in the BSL dictionary

developed using the same linguistic principles, the representation is more complex and is, we claim, best represented as a tree-structure.

3. Encoding sign semantics

The BSL dictionary typically gave several "glosses" to a sign that expressed its correspondence to English lexical meanings. We used the following heuristics in order to match a sign to an English key word:

- Key words were chosen from the glosses which seem to give the best coverage over all the other glosses, e.g., for entry 1609 in the dictionary: {abroad, exterior, external, foreign, foreigner, outdoor, outdoors, outside, overseas}, the word "external" was chosen;
- Key words which could be ambiguous in English but where the same ambiguity does not appear to exist in the sign were avoided, e.g., for 969: {fizzy drink, lemonade, pop}, "lemonade" was selected in preference to "pop";
- We chose single key words from the lists of glosses in preference to creating new ones because of the danger of experimenter bias;
- Some signs only have gloss phrases and not key words, and so a new key word had to be found, e.g., 526{take advantage} was matched with "exploit".

We then generated a high-dimensional co-occurrence based vector to represent the meaning of the sign using the methods describe in Bullinaria & Levy [11] We and others [16, 17] have demonstrated that this kind of high dimensional vector is highly effective at capturing lexical semantic "distance" or relatedness as measured by tasks such as picking out the most synonymous match for a word in a vocabulary test.

To create the semantic vectors for each target gloss word, we collected co-occurrence counts for the target with any other of the most frequent 50,000 words occurring within a window of plus-or-minus one word in the 90 million word text-based section of the British National Corpus [10]. The co-occurrence counts were then converted into positive pointwise mutual information values between the target word and each context word giving a 50,000 element semantic vector for each word of interest. We have demonstrated that these methods and parameters are highly effective in a number of different evaluation measures [11].

It was necessary to use English co-occurrence statistics to generate the semantic vectors as we know of no sufficiently large transcribed corpus of BSL that would have allowed us to measure the necessary co-occurrence statistics. We make the reasonable assumption that the mutual distances for the vectors for the concepts of, say, cat, dog and cot derived from a corpus of English will correlate highly with the usages of the corresponding BSL signs.

4. Distance metrics

The choice of a distance metric for previous work using relatively simple phonetic feature vectors for monomorphemic spoken words has been reasonably straightforward [8]. There was a greater possible choice of distance metrics for our tree-based representation based on the BSL dictionary. Previous work in ASL has examined how different aspects of sign similarity are perceived. For example, Lane, *et al* [18] examined the distinctive features for ASL handshapes that accounted for confusion matrices for identification in visual noise; Stungis [19] showed that both signers and non-signers grouped handshapes similarly; Poizner and Lane [20] reported a similar analysis for body location in ASL. Hildebrandt and Corina [21] demonstrated that ASL sign movement and location parameters were used by native signers, non-signers and late-learning signers but handshape was only relatively highly distinctive for late-learning signers. In order to cover the whole BSL dictionary, we included every feature necessary to distinguish the words in the dictionary. It is unlikely that all the features should have equal weight in a psychologically realistic distance metric. Without extensive empirical data to estimate such parameters, we chose a number of simple similarity measures that left the different features unweighted and explored other major components of the way in which two gestural forms could vary with respect to feature values and the branch paths required to achieve the feature values:

- **Feature only binary cosine similarity**: features were compared no matter where in a sign they appeared;
- **Branch-feature exact match binary cosine similarity**: similarities between two signs are registered wherever a feature is found that also occurs at the same point in execution of both signs;
- **Branch only binary cosine similarity**: branches were compared no matter what features were present – this is a measure of structural similarity;
- **Step feature and branch distance**: this is a simplified version of edit distance, allowing both feature and branch similarities to be registered separately as well as in combination. A normalized sum of all matching branch-feature pairs is computed where costs of one are added for cases where a feature matches and a branch doesn't or vice versa, a cost of four is added for each branch-feature pair that needs to be substituted and a cost of two is added for each branch-feature pair that is added or deleted.

As described and validated in Bullinaria & Levy [11], we used a vector (real-valued) cosine measure to measure semantic similarity.

5. Results

Correlation coefficients were calculated between corresponding form- and meaning similarities (e.g., between the form similarities between "cat" and "cot", "cat" and "dog", "cot" and "dog" and the semantic similarities between "cat" and "cot", "cat" and "dog", "cot" and "dog"). The coefficients served as measures of form-meaning systematicity and, following Shillcock *et al* [8], we employed Monte Carlo techniques to estimate the probabilities that the coefficients would have been measured that high by chance.

Table 1. Correlation results with 1-tailed p-values and z-scores for 10,000 Monte Carlo iteration for Pearson's and 100 iterations for Spearman's.

Form distance	Pearson's r	p-value / z-score	Spearman's rho	p-value / z-score
feature only	0.0365	<0.0001 / 12.7	0.033	<0.01 / 3.9
branch-feature exact	0.0370	<0.0001 / 11.7	0.040	<0.01 / 4.3
branch only	0.0255	<0.0001 / 6.8	0.038	<0.01 / 4.0
feature and branch step	0.0430	<0.0001 / 11.3	0.067	<0.01 / 6.1

Table 1 shows that we obtained small but statistically significant correlations for all of our distance measures. The values are comparable to the results obtained by Shillcock *et al.* They demonstrate that across the dictionary as a whole, there is only a small degree of systematicity between form and meaning similarities. The "feature only" result demonstrates that the feature values themselves without any measure of structure are sufficient to generate the small degree of correlation that we can detect. Matching branch structure alone gives the smallest correlation. However, this demonstrates that there is a significant degree of systematicity measurable from just the structural components of the tree representation – differences such as whether one or both hands are used, the order and number of movements and whether non-manual features are present. Using both branch structure and feature values generates the highest correlations with the "feature and branch step" method giving the highest results. This method is a form of edit distance and its "cost" weights could, in future, be altered to fit psychological data on sign similarity.

Table 1 allows us to compare the results for the Pearson correlation coefficient with those for the non-parametric Spearman's coefficient that makes less assumptions about the distributions. It is apparent that the results are comparable to Pearson's yielding significant and in most cases higher degrees of correlation.

We tested the stability of the correlation values by removing the 59 regional signs (only used in parts of the UK) leaving 1486 signs. This made only a small difference to the values. A Monte Carlo analysis where 1000 random selections of 59 signs were excluded demonstrated that the correlation values were stable.

We have described above how we developed a gestural representation that was capable of covering almost the entire BSL dictionary. This allowed us to make use of as many signs as possible. Shillcock *et al* [8], and Tamariz [13] restricted their samples to monomorphemic spoken words. One of the reasons they did this was to make it easier to exclude morphological markers that would have inflated their estimates of systematicity. By choosing to include nearly all the signs in the BSL dictionary, we may have included more morphosyntactic markers than was the case for the spoken word analyses and this may have inflated our estimates. On the other hand, we have noted above that it is unlikely that all the features in a complex sign carry equal psychological weight in the perception of sign similarity, so our estimates may have been distorted by our uniform form distance measures. Another possible way in which we may be losing some statistical power is that by comparing all the signs, we are measuring form similarities between words of different complexities and lengths. It is difficult for our distance measure to reflect the similarity between a simple sign made using one hand gesture and a complex one using two. Future analyses might include dividing the dictionary into signs made with a single hand (single dez), signs made with one stationary hand and one moving hand (manual tab) and signs made with two moving hands (double dez).

6. Discussion

We have developed a computational representation for BSL based on the notation used in [14] and explored several plausible distance/similarity measures. Using this gestural representation together with the semantic representation from [11], we have demonstrated that the distributional methods developed by Shillcock *et al* [8] can be successfully applied to a sign language. Our results suggest that the overall level of systematicity between form- and meaning-similarity in BSL is low and apparently comparable to that exhibited by spoken English.

Our results do not contradict any intuitions of a relatively high degree of absolute iconicity in BSL but rather that any such iconicity appears not to be used systematically in order to make similar signs mean similar things. Although there is no apparent absolute iconicity in many BSL signs, there are some examples of sign forms that appear to directly carry semantic information. For example, the BSL sign for apple appears to be a grasping of a small object held under the chin with a sharp twist of the hand. The BSL sign for banana uses two hands to trace out the curved shape of a banana. However, the semantic relatedness of the concepts of "apple" and "banana" does not appear to be reflected in similarities between their forms. Our distributional methods have provided a starting point for a quantitative analysis of these issues across the lexicon or significant fragments of it such as the established signs in the BSL dictionary.

There have been several recent discussions about the cognitive computational implications of the degree of systematicity and arbitrariness between form and meaning in spoken language [22, 23]. We hope that the work reported here on aspects of a particular sign language lexicon will extend the range of theoretical argument in this field so that sign languages can be considered alongside spoken languages. The tools that we have developed here may also be useful in discussing the plausibility for spoken language having evolved via a gestural stage [24, 25].

Acknowledgments

We gratefully acknowledge John Bullinaria's help with producing the semantic vectors as well as his helpful suggestions. This work was supported by the School of Human & Life Sciences, Roehampton University.

References

1. de Saussure, *Cours de linguistique générale*. Payot, Paris (1916).
2. Plaut, D. & Shallice, T. *Cognitive Neuropsychology*, **10**, 377-500 (1993).
3. Gasser, M., Sethuraman, N. and Hockema, S. Ch 1 in Rice, S. and Newman, J. (eds) *Experimental and Empirical Methods*. CSLI Press (2005).
4. Taub, S. F. *Language From the Body*, Cambridge University Press (2001).
5. Brennan, M. *The Visual World of British Sign Language: An Introduction* in British Deaf Association, *Dictionary of BSL* (1992).
6. Hickok, G., Bellugi, U. and Klima, K. *Nature*, **381**, 699-702 (1996).
7. Shillcock, R. C., Kirby, S., McDonald, S., and Brew, C. *Proceedings of the 2001 Conference on Disfluency in Spontaneous Speech*, 53-56 (2001).
8. Shillcock, R. C., Kirby, S., McDonald, S., and Brew, C. (under review).

9. Black, A., Taylor, P. and Caley, R. http://www.cstr.ed.ac.uk/projects/festival/manual/ (1999).
10. Aston, G. and Burnard, L. *The BNC Handbook: Exploring the British National Corpus with SARA*. Edinburgh University Press (1998).
11. Bullinaria, J. A. and Levy, J. P., *Behavior Research Methods*, in press.
12. Tamariz, M. and Shillcock, R. C. *Proceedings of the 21st Annual Meeting of the Cognitive Science Society*, 1058-63 (2001).
13. Tamariz, M. *Exploring the Adaptive Structure of the Mental Lexicon*, PhD thesis, University of Edinburgh (2005).
14. British Deaf Association, *Dictionary of British Sign Language/English*. London: Faber and Faber (1992).
15. Stokoe, W. C., Casterline, D. and Croneberg, C. *A dictionary of American Sign Language on linguistic principles*, Washington DC: Gallaudet University Press (1965).
16. Lund, K. and Burgess, C. *Behavior Research Methods, Instruments & Computers*, **28**, 203-208 (1996).
17. Landauer, T. K. and Dumais, S. T., *Psychological Review*, **104**, 211-240 (1997).
18. Lane, H., Boyes-Braem, P., and Bellugi, U. *Cognitive Psychology*, **8**, 263-289 (1976).
19. Stungis, J. *Perception & Psychophysics*, **29**, 261-276.
20. Poizner, H and Lane, H. In Siple, P. (Ed) *Understanding language through sign language research*, San Diego, CA: Academic Press, 271-287 (1978).
21. Hildebrandt, U. and Corina, D., *Language and Cognitive Processes*, **17(6)**, 593-612 (2002).
22. Gasser, M. *Proceedings of the 26th Annual Meeting of the Cognitive Science Society*, Hillsdale, NJ: LEA, 434-439 (2004).
23. Monaghan, P. and Christiansen, M.H. In Cangelosi, A., Smith, A.D.M., and Smith, K. (Eds.), *The evolution of language*, 430-432. New Jersey: World Scientific (2006).
24. Gentilucci, M. & Corballis, M. C. *Neuroscience & Biobehavioral Reviews*, **30**, 949-960 (2006).
25. Arbib, M. A. in Christiansen, M. A. and Kirby, S. (Eds) *Language Evolution*, Oxford University Press, 182-200 (2003).

Section III

Categorization and Visual Perception

TRANSIENT ATTENTIONAL ENHANCEMENT DURING THE ATTENTIONAL BLINK: EEG CORRELATES OF THE ST^2 MODEL

SRIVAS CHENNU*, PATRICK CRASTON, BRAD WYBLE AND HOWARD BOWMAN

Centre for Cognitive Neuroscience and Cognitive Systems,
University of Kent, Canterbury CT2 7NF, United Kingdom
**Email: sc315@kent.ac.uk*
Web: www.cs.kent.ac.uk/people/rpg/sc315

There has been considerable recent interest in the identification of neural correlates of the Attentional Blink (AB), and the development of neurally explicit computational models. A prominent example is the Simultaneous Type Serial Token (ST^2) model, which suggests that when the visual system detects a task-relevant item, a spatially specific Transient Attentional Enhancement (TAE), called *the blaster*, is triggered. This paper reports on our investigations into EEG activity during the AB, and a hypothesized correlation between the blaster and the N2pc ERP component. Specifically, we demonstrate that the temporal firing pattern of the blaster in the model matches the N2pc component in human ERP recordings, for targets that are seen and missed inside and outside the attentional blink window. Such a correlation between a computational account of the AB and ERP data provides useful insights into the processes underlying selectivity in temporal attention.

Keywords: Attentional Blink; ST^2 Model; Transient Attentional Enhancement; ERP; N2pc.

1. Introduction

The Attentional Blink (AB)[1,2] is a well studied temporal attention phenomenon, particularly suitable for investigating the nature and limits of conscious perception. The AB employs Rapid Serial Visual Presentation (RSVP), in which a sequence of items is presented at the same spatial location at a rate of around 10 items per second, with each item rapidly replacing the previous one. At such speeds, the items presented, some of which are targets to be detected and others distractors, yield only fleeting mental representations. The AB describes the finding that the detection of a second target (T2) following a correctly identified first target (T1) is significantly impaired if T2 follows T1 within 200–600 ms. There has been considerable recent interest in the identification of neural correlates of the AB, and the development of neurally explicit models, a prominent one being the Simultaneous Type Serial Token (ST^2) model.[3] In addition to incorporating a computationally explicit account of visual processing, attentional selection and working memory (WM) encoding, the model proposes the episodic distinctiveness hypothesis, i.e., that the AB reflects the visual system attempting to allocate unique episodic contexts to targets. Importantly, it suggests that when the visual system detects an item that may be task relevant, a spatially specific Transient Attentional Enhancement (TAE), called the *blaster* is triggered. For a fleeting stimulus, the contribution of this enhancement is critical in enabling it to be encoded into WM.

This paper reports on our investigations into EEG activity during the AB, and a hypothesized correlation between the blaster and the N2pc ERP component. The N2pc describes a negative deflection of the ERP at around 200–300 ms after the presentation of a laterally offset target, and is most strongly visible at parietal electrodes contralateral to the position of the target. Previous research has associated the N2pc with the selection of a target in the presence of competing distractors,[4] and as an indicator of the moment-to-moment deployment of attention.[5] In this paper, we describe an experiment employing a dual stream AB paradigm designed to record lateralized electrophysiological (EEG) activity during RSVP. We discuss experimental data that, when examined in light of predictions from the ST^2 model, suggests a correlation between the triggering of the blaster and the manifestation of the N2pc. Specifically, we demonstrate that the temporal firing pattern of the blaster in the ST^2 model matches the N2pc component in human ERP recordings, for targets that are seen and missed inside and outside the attentional blink window. This connection supports the hypothesis that the N2pc reflects the selective attentional enhancement that is embodied by the blaster.

This paper is organized as follows. Section 2 begins with a brief description of the ST^2 model, specifically focusing on the blaster and its influence on model dynamics. Section 3 shifts focus to a discussion of how blaster activation traces can be meaningfully compared to the N2pc. Section 4 describes the dual stream experiment designed to record the N2pc and elaborates on the correlational evidence gathered to connect it to the blaster.

Section 5 concludes with a general discussion and proposes a potentially reciprocal relationship between cognitive modelling and ERPs.

2. The ST^2 Model

The Simultaneous Type Serial Token theory of temporal attention, developed and explained in-depth in Ref. 3, provides a specific account of temporal information processing in the visual system and the formation of WM representations. Its connectionist realization, the neural-ST^2 model as depicted in a simplified form in Figure 1, implements its principles in a fixed-weight neural network. The architecture of the model can be divided into two stages of processing, both of which interact with the *blaster*, which provides a brief but powerful attentional enhancement in response to the occurrence of targets. This two-stage design, inspired by Ref. 1, supports the hypothesis that the AB is characterized by a late-stage processing bottleneck as described below.

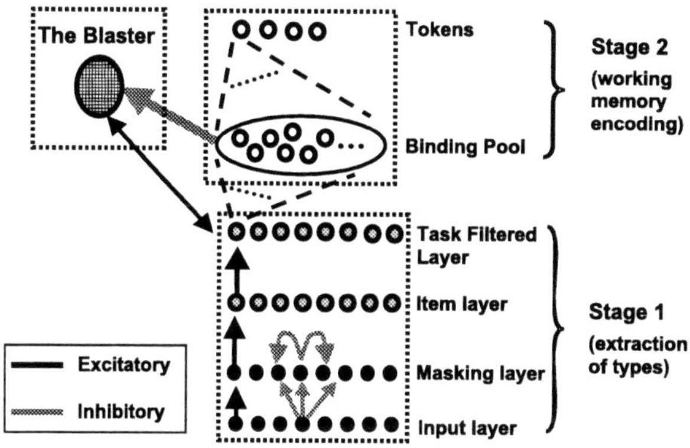

Fig. 1. The Neural-ST^2 Model

Stage 1 abstractly models early visual processing common to all stimuli, including visual masking and semantic categorization. Stimuli are processed in parallel and filtered based on task specific salience as they pass through Stage 1. Its pipelined design implies that the representation of a given item in Stage 1 is cascaded across multiple layers at any time.[6] In its final layer, Stage 1 generates rapidly decaying representations of featural characteristics of items, which in this framework are called *types*.[7] Only targets trigger the blaster, which provides additional excitation to support their successful encoding into WM.

Stage 2 creates and maintains durable representations of targets in WM. Targets at the end of Stage 1 pass into Stage 2 by binding their type to a *token*. Unlike in Stage 1, this *tokenization* process is strictly sequential, and attempts to associate distinct episodic contexts to targets. This serialization of target encoding is enforced by actively inhibiting the blaster and preventing it from refiring during ongoing tokenization.

The blaster is triggered by the detection of a target at the end of Stage 1, which in turn elevates activation levels across the final layers of Stage 1. This transient attentional enhancement provided by the blaster lasts for a brief (150ms) window of time following target detection. The availability of this attentional boost from the blaster is essential for most targets because of the fleeting representations generated in RSVP, which usually do not have sufficient bottom up strength to initiate WM encoding. A key feature of the blaster is that it fires only once per tokenization. Once a target (T1) triggers the blaster and the process of binding is initiated, it is held offline by inhibition from Stage 2 till the process is complete. This inhibition collapses the attractor previously set up, and attempts to associate distinct episodic contexts to targets. It does so by preventing a second target (T2) in close temporal proximity to T1 from interfering with its tokenization. T2 must "wait" till the T1 tokenization is complete for the blaster to fire again. This implies that only T2s with strong bottom up strength have enough activation to "outlive" T1 tokenization. It is this mechanism, embodying the episodic distinctiveness hypothesis, that enables the ST^2 model to simulate the attentional blink.

Model Performance Figure 2 compares the performance of the ST^2 model to human behavioural data relating to the AB.[1] It focus on performance in the basic AB scenario, when T1 is followed by a blank, and when T2 is at the end of the RSVP stream. In addition to these scenarios, the model reproduces a broad spectrum of AB data. See Ref. 3 for a detailed comparison.

3. Connecting Modelling and ERPs

Computational modelling in the context of the AB has tended to focus on replicating behavioural data. The ST^2 model itself was conceptualized and designed to simulate patterns of behavioural data collected in AB experiments. In this paper, we look at whether the connection between modelling and experimentation can be extended from the behavioural to the electro-

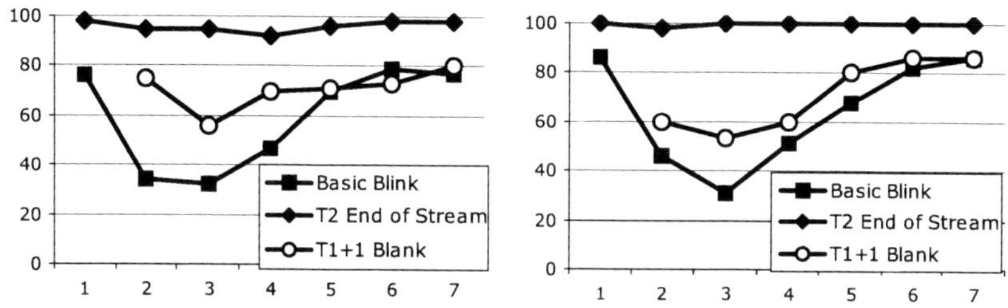

(a) Human (b) Model

Fig. 2. Comparison of the ST^2 model to behavioural data

physiological domain. Although successful replication of a broad spectrum of behavioural data relating to the AB is convincing validation of the ST^2 model, being able to relate ERPs to specific parts of the model allows for a more fine-grained verification. To make these correlations, we build upon the correspondence between specific parts of the model and successive stages of temporal visual processing that generate ERP components. Though the approach used in this paper to connect modelling and electrophysiology is exploratory, it allows for the use of another dimension of experimental evidence for validating specific processing stages proposed in the model.

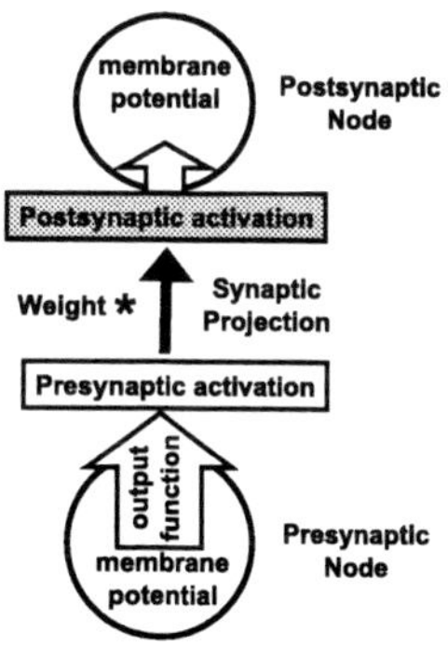

Fig. 3. A pair of model neurons

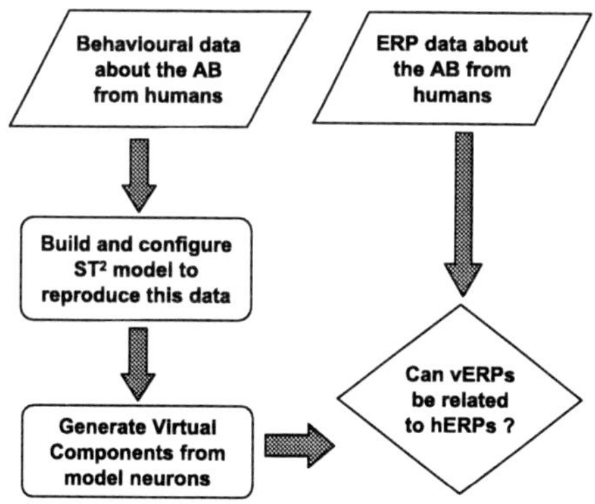

Fig. 4. Can vERPs be related to hERPs?

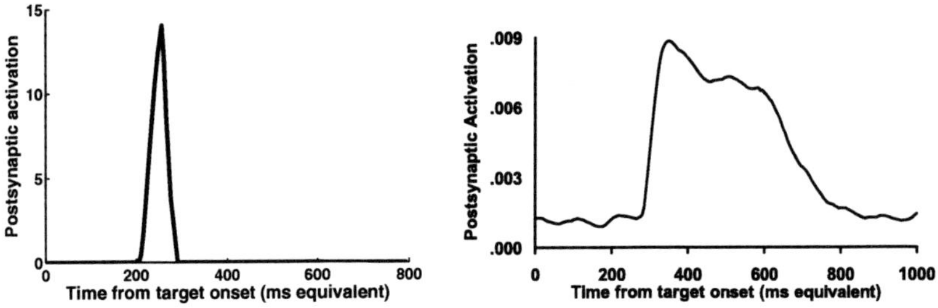

(a) Blaster Output

(b) Virtual P3

Fig. 5. Virtual ERPs from the the ST^2 model

Virtual ERPs from ST^2 In order to compare activation dynamics of the neural-ST^2 model to human ERPs, we generate *virtual ERPs* from traces produced by specific processing layers recorded during simulations. Virtual ERPs are thus defined as grand averages of neural activation, summing over all possible combinations of target strengths, and represent the typical pattern of activity set up in the model when a particular experimental condition is simulated. Figure 3 depicts a pair of nodes in the model, connected by a weight that represents an excitatory synaptic projection. The membrane potential of the presynaptic node feeds into its output function and the resulting presynaptic activation contributes to the postsynaptic membrane potential after being multiplied by the intervening weight. This method to generate virtual ERPs is intended to best approximate the mechanism assumed to be responsible for the generation of human ERPs. Electrical activity observed at the scalp is thought to reflect the summation of the postsynaptic potentials generated at a large number of spatially aligned pyramidal neurons in the cortex,[8] which are known to release the excitatory neurotransmitter glutamate. Analogously, a virtual ERP comparable to a human ERP is the average excitatory postsynaptic potential recorded across functionally equivalent layers in the model. The functional role of the layers chosen to generate a particular virtual ERP is hypothesized to correspond to the neural processing that generates the corresponding human ERP. Though external factors like scalp distortion have not been considered in generating virtual ERPs, we think that being able to make qualitative comparisons and predictions about human ERPs using virtual ERPs can contribute to the process of cognitive modelling.

Figure 5 shows sample virtual ERPs generated from the ST^2 model. Figure 5(a) depicts the burst of activation generated by the blaster in response to a single target. Figure 5(b) depicts the virtual P3 generated during the encoding of that target, calculated by averaging activation across those layers in the model that simulate target selection and WM encoding by associating the type of the target with a token.

Given this definition of virtual ERPs, the question we pose is summed up in the diagram depicted in Figure 4. Empirical research on the AB has produced a substantial dataset of behavioural and ERP data. The ST^2 model is used to generate virtual ERPs from model layers that simulate distinct cognitive functions. In juxtaposing human ERP data with these virtual ERPs, we are interested in investigating whether being able to find correlations between them leads to useful insights into both the architecture of the model and the human ERPs themselves.

3.1. *The Blaster and the N2pc*

We now state the key hypothesis of this paper; that there is a correlation between the blaster component in the ST^2 model and the N2pc ERP component. The blaster, as has been described thus far, is responsible for providing the transient attentional burst necessary for encoding targets into WM. The N2pc ERP component, on the other hand, is a well-studied negative deflection occurring in the ERP waveform 200–300 ms after the onset of a salient stimulus. The N2pc is a lateralized component, in that it is larger at parietal electrodes contralateral to the stimulus position, and is usually plotted as a difference waveform obtained by subtracting the ipsilateral electrode from the contralateral one. The N2pc has been observed in spatial visual search experiments[4,5,9] and in RSVP paradigms.[10–12] It is thought to reflect the locus of visual spatial attention, and the tracking of the instantaneous deployment of attention.

This definition of the N2pc has similarities with the functional effect of the blaster. The preliminary hypothesis that we put forward is that the N2pc corresponds to the firing of the blaster in response to targets. In order to verify this hypothesis, we attempt to connect the firing pattern of the blaster to the relative strength of the N2pc in different experimental conditions. A key prediction from this hypothesis is that the blaster firing pattern, and consequently the N2pc amplitude, are fundamentally different for targets "inside" (T2 at lag 3) and "outside" (T1) the blink window. As will be discussed further in Section 4.2, all T1s get the benefit of the blaster and elicit a similar N2pc, irrespective of whether or not they are consciously reported. This is in contrast to T2s occurring during the blink, which are available for conscious report only if they get the benefit of the blaster.

Hence, seen T2s elicit a larger N2pc than missed ones. In order to test this hypothesis, we conducted an EEG experiment employing a dual stream AB paradigm designed to record the N2pc. The details of the experiment and the results obtained are the topic of the next section.

4. Dual Stream Experiment

We now describe an experiment employing a dual stream RSVP paradigm designed to produce the attentional blink and simultaneously record EEG activity.

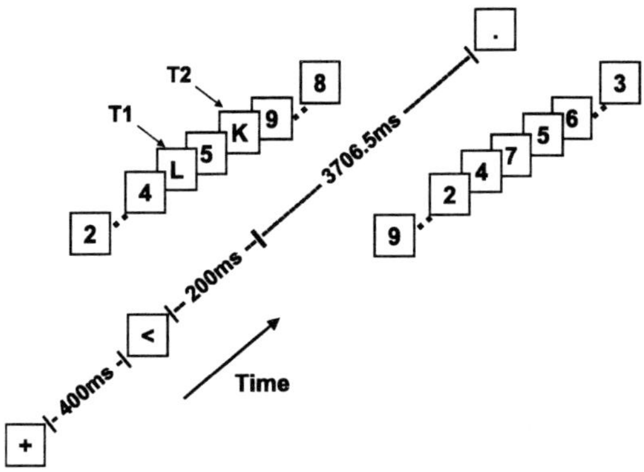

Fig. 6. A sample trial in the dual stream RSVP experiment

4.1. *Design*

Our experimental paradigm employed a pair of laterally presented RSVP streams consisting of letters and digits, where letters were targets to be identified, and digits were distractor stimuli. Figure 6 depicts the timeline of a sample trial. In each trial, 35 stimuli at 105.9ms SOA were presented in each of two RSVP streams at either side of fixation. The actual streams were preceded by a central fixation cross. This cross was replaced by an arrow after 400ms, indicating one of the two RSVP streams in which two target letters would appear. Participants were instructed to direct their covert attention to the indicated stream while continuing to fixate their gaze centrally. The streams began 200ms after the arrow and lasted for 3.706s. Immediately after the end of the streams, the central arrow turned into either a dot or a comma, chosen randomly, which stayed on for 105.9ms.

After the end of the trial, participants reported the identity of any targets they saw, and a dot or a comma for the last item. This additional task was included to ensure the participants fixated centrally throughout the presentation of the streams. Each participant was presented with 4 blocks of 100 trials each. A given trial could contain either 0 or 2 randomly chosen letter targets, appearing equally randomly in one of the two streams. The second target (T2) appeared at lag positions 1, 3 or 8 after the first one (T1).

Participants Experimental data from 14 university students (mean age 22.4, SD 2.9; 6 female; 13 right-handed) were included in the analysis. All were free from neurological disorders and had normal or corrected-to-normal vision.

EEG Recording Scalp EEG was recorded at 1000Hz (bandpass filtered at 0.25Hz–80Hz during recording) while participants performed the task, from 19 electrodes placed at standard 10/20 locations (Fp1, Fp2, Fz, F3, F4, F7, F8, Cz, C3, C4, C7, C8, Pz, P3, P4, P7, P8, O1, O2, T7 and T8). In addition, a bipolar EOG channel below and to the left of the left eye recorded eye movements, used to reject trials with artefacts. Recorded EEG data was referenced to a common average online and re-referenced to linked earlobes offline. An electrode at the left mastoid acted as ground.

EEG Analysis For each subject, EEG segments for conditions of interest were time-locked to the onset of the target and extracted from -200ms to 800ms with respect to target onset. For each such segment, direct current drift artefacts were removed using a DC de-trend procedure employing the first and last 100ms of each segment. The segments were baseline corrected to the 0–150ms interval following target onset that does not contain any lateralized ERP components, and then averaged together for each condition. The average additional activity elicited by the targets for a particular subject was then calculated by subtracting the contralateral parietal electrode from the ipsilateral one, either P7 or P8, in order to isolate the N2pc. For plotting the waveforms, they were low pass filtered at 25Hz for visual clarity.

4.2. *Results and Discussion*

This section compares human ERPs from the dual stream experiment and virtual ERPs from the ST^2 model, with regard to the hypothesis put forward in Section 3.1

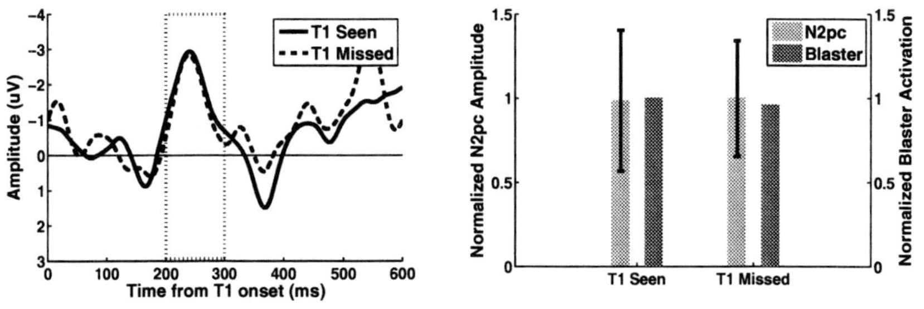

(a) N2pc for seen vs. missed T1s

(b) Blaster for seen vs. missed T1s

Fig. 7. Comparing the N2pc and the blaster for targets outside the blink

Targets outside the blink Figure 7(a) depicts the N2pc elicited by seen vs. missed T1s in the 200–300ms window following T1 onset. As can be clearly seen, the N2pc has similar amplitude, onset and offset for both conditions. This fact is reflected in the results of a repeated measures ANOVA that shows no significant effect of condition $(F(1, 13) = 0.001, MSE = 1.466, p = 0.98)$. This implies that T1s, i.e., targets presented outside the blink window, elicit a similar N2pc, irrespective of whether they are eventually seen or missed. To correlate this finding with the ST^2 model, Figure 7(b) compares the normalized postsynaptic activation of the blaster to the normalized N2pc amplitude for the same pair of conditions. The postsynaptic activation plotted here is averaged over all possible target strengths in the model. Two key observations can be made from this comparison: firstly, the amplitudes of the blaster for seen and missed T1s are similar. This can be understood by noting that in the ST^2 model, the blaster always fires for T1s as it is not inhibited by any ongoing tokenization. Hence T1s that are missed in the model are too weak to complete tokenization, despite the attentional enhancement provided by the blaster. Secondly, the average amplitude of blaster activation covaries with the N2pc for seen and missed T1s.

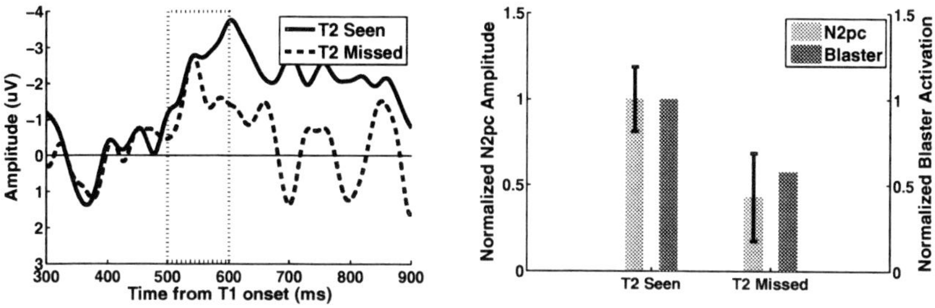

(a) N2pc for seen vs. missed T2s (b) Blaster for seen vs. missed T2s

Fig. 8. Comparing the N2pc and the blaster for targets inside the blink

Targets inside the blink Figure 8(a) depicts the N2pc elicited by seen vs. missed T2s at lag 3, i.e. well within the attentional blink, during the 200–300ms window following T2 onset. The human ERPs have been computed by averaging over only those trials in which T1 was seen. In contrast to the previous comparison, the N2pc for seen vs. missed T2s is markedly different. Specifically, it is larger in amplitude and area for seen T2s. An ANOVA supports these observations by suggesting a significant effect of condition ($F(1,13) = 8.133, MSE = 2.699, p = 0.013$). This variation in the N2pc is mirrored by the postsynaptic activation of the blaster, as depicted in Figure 8(b). The blaster is inhibited during the blink window and can fire again only after T1 tokenization has completed. Only those T2s at lag 3 that have enough activation left at the end of T1 tokenization manage to refire the blaster. Of these, some are nevertheless missed due to insufficient bottom-up strength. As a result, the average blaster activation for missed T2s at lag 3 is significantly lesser than that for seen T2s.

5. Conclusions

The comparisons between the blaster and the N2pc emphasize a key property of the ST^2 model of the attentional blink: the unavailability of the transient attentional enhancement for targets during the blink window. This *temporal spotlight* of attention provided by the blaster, though necessary for tokenization, is not sufficient. The ERP data shows that both seen and missed T1s elicit a similar N2pc, but the N2pc for seen T2s at lag 3 is significantly larger than that for missed T2s. Together, the N2pc data and the blaster activation traces point toward a key observation: the deployment of transient attentional enhancement is fundamentally different for targets inside and outside the blink.

Implications for Modelling and ERPs This paper also highlights the more general idea of connecting cognitive modelling and ERPs. Though the attempt to do so in this paper is a qualitative one, the results are encouraging. To a large extent, the computational detail of the ST^2 model has enabled us to generate virtual ERPs to compare and contrast with human ERPs. But this level of detail comes with implementation costs. As with any model required to reproduce a broad spectrum of data, there is a trade-off between modelling capability and design complexity, since attempting to model a broad spectrum of data tends to increase the computational requirements of the model.

In spite of these challenges, being able to connect model dynamics to human ERP data has reciprocal benefits. Firstly, generating ERPs from computational models allows us to make predictions about the pattern of variation in human ERPs that we can expect to see across experimental conditions. Models inspired by neurophysiologically plausible architectures can be used to theorize about and direct the effort to better understand the neural sources of ERPs in the brain. Reciprocally, ERP data can be used to comparatively evaluate competing theories of psychological phenomena.

References

1. M. Chun and M. Potter, *Journal of Experimental Psychology: Human Perception and Performance* **21**, 109 (1995).
2. J. Raymond, K. Shapiro and K. Arnell, *Journal of Experimental Psychology: Human Perception and Performance* **18**, 849 (1992).
3. H. Bowman and B. Wyble, *Psychological Review* **114(1)**, 38 (2007).
4. M. Eimer, *Electroencephalography and Clinical Neurophysiology* **99**, 225 (1996).
5. G. F. Woodman and S. J. Luck, *Journal of Experimental Psychology: Human Perception and Performance* **29**, 121 (2003).
6. J. L. McClelland, *Psychological Review* **86**, 287 (1979).
7. N. Kanwisher, *Journal of Experimental Psychology: Human Perception and Performance* **17**, 404 (1991).
8. S. J. Luck, *An Introduction to the Event-Related Potential Technique* (MIT Press, Cambridge, MA, 2005).
9. S. J. Luck, M. Girelli, M. T. McDermott and M. A. Ford, *Cognitive Psychology* **33**, 64 (1997).
10. P. Jolicoeur, P. Sessa, R. Dell'Acqua and N. Robitaille, *European Journal of Cognitive Psychology* **18(4)**, 560 (2006).
11. P. Jolicoeur, P. Sessa, R. Dell'Acqua and N. Robitaille, *Psychological Research* **70(6)**, 414 (2006).
12. R. Dell'Acqua, P. Sessa, P. Jolicoeur and N. Robitaille, *Psychophysiology* **43**, p. 394 (2006).

A DUAL-MEMORY MODEL OF CATEGORIZATION IN INFANCY[*]

GERT WESTERMANN

Department of Psychology, Oxford Brookes University
Oxford OX3 0BP, UK

DENIS MARESCHAL

Centre for Brain and Cognitive Development, School of Psychology, Birkbeck, University of London, London, WC1E 7HX, UK

Although cognitive neuroscientists have explored the neural basis of category learning in adults, there has been little if any investigation of how the unfolding categorization abilities of infants relate to the development of neural structures during the first two years of life. Here, we argue that category learning in early infancy can be explained through the interactions of two memory systems: a cortically-based long-term system, and a hippocampal short-term system. We suggest that the shift in categorization behavior observed in infant category learning from bottom-up empirical learning to learning that is strongly influenced by top-down prior knowledge reflects the gradual functional integration of these two systems. To test this hypothesis we describe a dual-memory connectionist model that implements interactions between the long term (neocortical) and short-term (hippocampal) networks.

1. Introduction

Forming categories is one of the most fundamental aspects of cognitive processing. Therefore, studying the development of this ability in infancy is vital for understanding the basics of cognitive processing as a whole. As infants cannot speak and communicate their knowledge verbally, non-linguistic methods of probing early categorization have been developed. However, different methods have yielded sometimes conflicting results on infants' categorization abilities [1], making it difficult to integrate them into an overall picture of category development. In this paper we suggest that the key to interpreting these conflicting results is to see them as based on different underlying representations in different memory systems, loaded differentially by

[*]This work was supported by the Royal Society, ESRC grant R00019326, and EC Framework 6 grant 516542 (NEST).

the different tasks used to assess categorization at different ages. In the rest of this chapter we first describe different methods for testing category formation in infancy. Next, we briefly review a previous approach to modeling infant categorization behavior and then describe a new neural network model that is based on the multiple memory view. We report several simulations with this model that examine the development of object representations and interactions between memory systems during categorization tasks.

1.1. *Category Formation in Infants*

One set of methodologies by which infant categorization has been studied relies on the fact that infants tend to show a preference for novel stimuli [2]. Studies exploiting this novelty preference usually employ a *familiarization* stage in which infants are shown a sequence of images of objects from one category (e.g., cats) on a computer screen. The time that the infants spend looking at each image is measured and is expected to decrease as infants become familiarized with the objects. This stage is followed by a *test* phase in which infants are shown novel stimuli from the familiarized category (e.g., a novel cat) and stimuli from a different category (e.g., a dog). Preference for the object from the different category can then be taken as evidence that the infants have formed a category representation that includes the novel category member (the novel cat) but excludes the object from the other category (the dog). Research based on this paradigm has provided evidence that infants under six months of age can form perceptual categories even of complex visual stimuli such as different animals and furniture items [1,3,4]. The level (global or basic) at which objects are categorized is dependent on the variability and distribution of information in the environment [5,6]. For example, in one study [7], 3-4 month olds were familiarized with cats and subsequently were shown to have formed a basic-level category representation of domestic cats that excluded birds, dogs, horses and tigers. Likewise, when familiarized with chairs, infants formed a basic-level category representation of chairs that excluded couches, beds and sofas. In a different study [8], 3-4 month olds were familiarized on different mammals, resulting in their forming of a global-level category representation of mammals that included novel mammals but excluded non-mammals such as birds and fish, as well as furniture. When familiarized with different furniture items, infants formed a global-level category representation of furniture that included novel furniture items but excluded mammals. It therefore seems that infants at 3-4 months can show categorization on different levels. However, it has been argued that even younger infants appear to form global distinctions only [9].

Other experimental paradigms are not based on novelty preference and do not involve a familiarization stage. For example, in the *generalized imitation* paradigm [10] infants are shown a simple action involving toy figures, such as giving a cup of drink to a dog. The infant is then encouraged to imitate this event with different toys, e.g., a different dog, a cat or a car. Category formation is inferred from observing to which novel objects the infants generalize the modeled action. Due to the absence of familiarization this kind of task is assumed to tap into the background knowledge that infants have acquired during their everyday experiences [11]. In this paradigm it has been found that global category distinctions (such as animals vs. vehicles) emerge first at around 7 months of age, whereas basic level distinctions (such as cats vs. dogs) do not appear until around 14 months of age [12].

The different experimental paradigms used in infant category formation have given rise to conflicting theories of the mechanisms underlying early categorization. According to one view [e.g., 13] early category formation is entirely based on the perceptual properties of observed objects. With increasing experience, interacting with objects, and the onset of language, representations gradually become enriched to transcend this purely perceptual information towards more abstract concepts. By contrast, a dual process view of category formation [e.g., 11] assumes two separate mechanisms for perceptual and conceptual categorization, respectively. According to this view the perceptual mechanism is operational from birth, and a conceptual mechanism develops in the second half of the first year of life. Category formation is then based on integrating the separate representations emerging from both mechanisms.

Whereas the results from preferential looking and generalized imitation studies seem to contradict each other, with infants in preferential looking studies showing categorical differentiation much earlier than in generalized imitation studies, a possible reconciliation can be suggested by highlighting the different task requirements in these paradigms. Preferential looking studies examine within-task on-line category formation and analyze looking behavior to infer category formation. Generalized imitation studies tap into background knowledge and require complex motor responses. Thus, we argue that these different studies can be construed as providing a set of collective insights into the development of a complex neuro-cognitive system that contains multiple category learning and category storage systems.

The existence of multiple memory and category learning systems is well established in the adult literature [14,15]. Simply speaking, the idea is that there is a division of labor between a fast learning system in the hippocampus and a slow learning cortically-based system. The hippocampus is responsible for the rapid learning of new information, whereas cortical representations develop more gradually and integrate new with previously learned knowledge. Here we suggest that this approach can also account for the unfolding categorization abilities in infants.

Although little is known so far about the development of memory systems in infancy, Nelson [16] has hypothesized that novelty preference in infants relies on a hippocampal *pre-explicit* memory system that is functional from shortly after birth. According to Nelson, *explicit* memory becomes functional only after 6 months of age. It is based on the hippocampus as well as on cortical areas such as inferotemporal cortex and entorhinal cortex. For explicit memory to become fully functional it is thus necessary to develop the entire hippocampus, the relevant cortical areas, as well as connection between hippocampus and cortex. Furthermore, complex tasks such as deferred imitation appear to rely on interactions between the two memory system, involving the hippocampus as well as occipital, premotor, left inferior prefrontal and frontal cortices. From this perspective it is clear that categorization tasks relying on cortical representations will develop later than those relying on preferential looking.

1.2. *Previous Models of Infant Categorization*

Previous models of infant categorization have often employed auto-encoder neural networks [5,17,18]. These are simple 3-layer backpropagation models in which input and target are the same, that is, the model learns to reproduce its input on the output side. In most auto-encoders the hidden layer is smaller than the input and output layers, forcing the model to extract regularities from the input in order to reproduce it. The rationale behind using these networks for modeling infant categorization is that the network error can be linked to infant looking time. One theory of infant novelty preference is that when infants look at an object they gradually build up an internal representation of this object, and looking continues until internal representation and object match [19]. The more unusual an object is, the longer it will take to build this representation, and the longer the object will be fixated. Likewise, in the auto-encoder, successive weight adaptations lead to a match between an input (the object) and output (the infant's internal representation of the object). A higher output error requires more adaptation steps and thus can be likened to an infant's longer looking time.

While simple auto-encoder models have been successful in accounting for different results from infant looking time studies, they do not take prior knowledge into consideration. Thus, they can only be applied in simulations of within-task category formation that involves familiarization and does not take into account the infant's background knowledge. In Nelson's framework of memory development [16], simple auto-encoders implement pre-explicit memory only. However, a comprehensive understanding of infant categorization requires that the mechanisms of both within-task and long-term category formation are integrated into a single system. Furthermore, an infant's background knowledge can have an effect even on within-task category formation. For example, one study [20] showed that whether infants had cats and dogs at home affected their categorization of cats and dogs in an experimental setting. In another study, Quinn and Eimas [21] argued that young infants' categorization of humans versus non-human animals in a visual preference experiment is affected by differential experience with members from the two classes that occurs prior to the experiment.

The model described here extends previous categorization models by investigating the unfolding interactions between different memory systems through development. The model is loosely linked to the hippocampal/cortical memory systems in the brain. It therefore allows for the investigation of the development of category representations in long-term memory on the basis of experience with the world as opposed to laboratory-based experiments only. By modeling interactions between both memory systems it is also possible to examine the role of previously acquired knowledge on performance in a laboratory task.

2. An Integrated Model of Categorization in Infancy

The model (Fig. 1) consists of two linked auto-encoder networks, one of which represents the earliest, pre-explicit memory system based on the hippocampus, and the other represents later memory systems that are largely cortically based (see [14]). In line with theories of adult memory, the 'hippocampal' system is characterized by rapid learning (high learning rate) with susceptibility to interference (catastrophic forgetting), and the 'cortical' system by slower learning (low learning rate).[†] Unidirectional links between the two components implement interactions between the two memory systems.

Training of the model worked as follows: an input was presented and activation was propagated to the hidden layers. Activation then was cycled back and forth between the hidden layers (which were updating each other's activations) until a stable state was reached. Activation was then propagated to the output layers. All weights were adjusted with the backpropagation algorithm at each stimulus presentation.

[†] This model is not intended as a neuronal model. Thus, we are not arguing that there is a close relationship between hippocampus and cortex and the auto-encoder networks, but we will adopt the terms 'hippocampal network' and 'cortical network' for ease of expression. See [14] for a similar terminology.

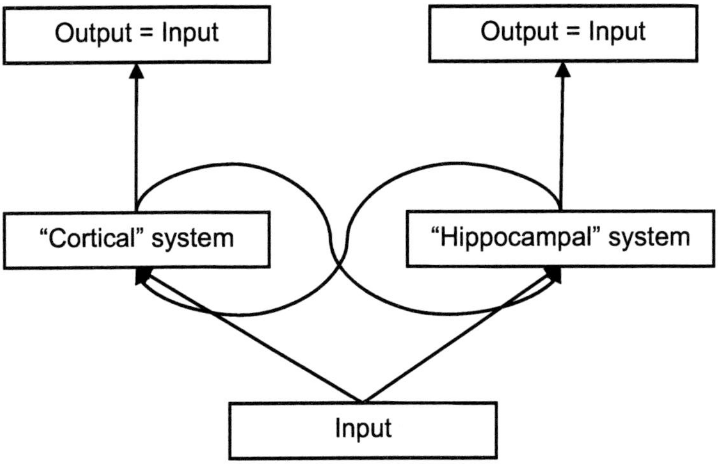

Figure 1: The architecture of the dual memory categorization model.

In the simulations reported here, the network parameters were as follows: learning rate in the hippocampal model: 0.5; learning rate in the cortical model: 0.001; learning rates of the lateral connections between the networks: 0.01; momentum for all weights: 0.4; size of each hidden layer: 15.

2.1. *Data*

Photographs of objects from 19 different basic-level categories were encoded according to a number of perceptual features. These were *maximal height, minimal height, maximal width, minimal width, minimal width of base, number of protrusions, maximal length/width of left, right, lower and upper protrusion, minimal width of lower protrusion, texture, eye separation, face length, and face width*. These features comprise both general (geometric) and object specific (facial) characteristics. Feature values were scaled between 0 and 1.

The object categories were chairs, sofas, tables, beds, drawers, horses, giraffes, birds, rabbits, squirrels, elephants, deer, fishes, cars, males and females. They fell into four global-level categories (furniture, animals, vehicles, humans) all previously used to test infant category formation [7] and varied in their within-category perceptual similarities. Each category consisted of ten exemplars (for some categories additional exemplars were created by interpolating the representations of the measured photos). For each category a prototype was generated by averaging the representations of all members of that category.

The model was trained in two different ways. *Familiarization Training* was meant to replicate the experience of an infant in a laboratory setting: the model was first familiarized on a set of stimuli until criterion was reached and was then tested on test stimuli. *Background Training* aimed to capture the experience of an infant with the world by training on random stimuli for random amounts of time before switching to a new random stimulus.

2.2. *Results*

2.2.1. *Development of long-term representations*

In a first experiment the hidden representations developed by the cortical component of the model were explored. This was interesting for three reasons: the first was that a neural network that is trained on a sequence of patterns can be subject to catastrophic forgetting [22]: when a new pattern is learned, the representations for previous patterns can be overridden and the learned mapping for these patterns is lost. Here we wanted to investigate if catastrophic forgetting can be avoided through the interactions between the fast and slow components in the model. The second reason for studying the hidden representations of the model was that there was overlap between the perceptual representations of patterns from different categories. This experiment could therefore show how the model clustered exemplars based on perceptual information alone. The third reason was the assumption made in the model that complex responses such as generalized imitation depend on cortical representations. In these tasks infants show a global-to-basic development of categories, and this could be explained by a similar development of cortical category representations.

The model was trained in background training mode for 10 epochs, that is, each of the 190 training exemplars were presented to the model in random order, and for random lengths of time (between 1 and 1000 weight updates), for 10 times. To assess the effect of interactions between the hippocampal and cortical components, the model was trained ten times with lateral connections between the hidden layers, and ten times without. Results were thus averaged over 10 runs of each model.

Figure 2 shows the development of the hidden representations for category prototypes in the cortical component at four points in training, after exposure to 5, 50, 100 and 1500 stimuli. Over training the representations become more distinct, with global level categories (humans, furniture, animals) already well separated after 100 stimuli, and basic level subdivisions (e.g., male-female;

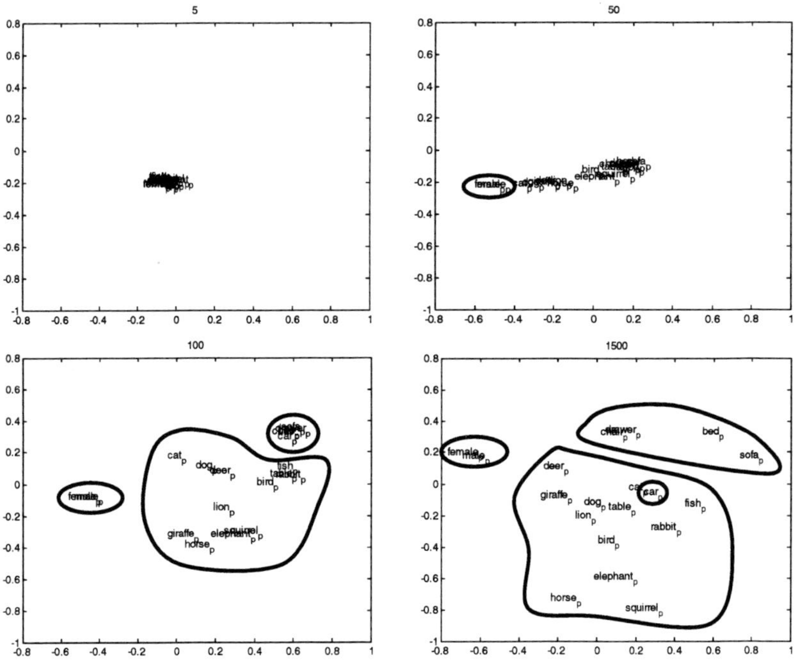

Figure 2: Development of the hidden representations of category prototypes (subscript p) in the cortical component of the model. Representations are plotted in terms of their first two principal components.

bed-sofa) visible after 1500 stimuli.[‡] This result provides an explanation for the global-to-basic shift in category development that is found in non-familiarization tasks.

The spread of representations was compared for 10 models each trained with and without lateral connections between the cortical and hippocampal components. For the models trained with lateral interactions, the average pairwise distance between cortical hidden representations for all stimuli was 0.74 and for models trained without lateral connections (i.e., with no interaction between memory systems) this distance was 0.38, which was significantly smaller ($F(9) = 34.66$, $p<.0001$). Thus, hippocampal input to cortex in addition to direct sensory input led to an increased differentiation of cortical category representations, indicating that interactions between memory systems support

‡ Because of the relatively limited set of image views used in the training set (e.g., only canonical views of tables), inevitable idiosyncrasies in object representations arose (e.g., the close similarlity between dogs and tables in the current simulations). A more representative sample of object views would presumably avoid these spurious similarities.

differentiation of long-term representations and reduce interference between objects.

2.2.2. *The effect of background knowledge on familiarization*

We investigated the effect of background knowledge of different categories on the time required to reach familiarization criterion in hippocampus-based familiarization tasks. For background knowledge to have an effect, cortical representations must alter hippocampal representations that arise from perceptual experience with objects during a familiarization task.

To investigate this question, a simulation was carried out in which dogs were used as familiarization items, and prior background training was on either all items except dogs, or on all items except animals. The prediction was that familiarization times for dogs would reduce more when the background training data contained animals than when it did not. In the Animals condition the model was trained on all stimuli, excluding dogs but including other animals, for 10 epochs (1800 exemplars). In the No-Animals condition, the model was trained on all stimuli excluding animals for 22 epochs (1760 exemplars; training was on more epochs than in the Animals condition to ensure an approximately equal number of exemplars).

The results of this simulation are displayed in Figure 3A. Background training (experience with the world) led to a significant decrease in familiarization time. Familiarization time (stimulus presentations necessary to reach error criterion) to dogs was significantly shorter when the model had previous experience with animals than when it had not been trained on background knowledge. However, the specific type of background knowledge also made a difference: models that had prior experience with animals also familiarized significantly faster than those that only had experience with humans, vehicles and furniture ($F(99)= 3.6967$, $p<.001$). This result indicates that prior learning from experience with the environment can affect an infant's performance in a familiarization experiment in the lab.

The effect of prior knowledge on familiarization was further explored for cases in which an infant has experience with members of the specific category that is then tested in the lab. For this simulation, one random dog exemplar was removed from the stimulus set. Then, the model was background-trained either on the remaining 9 dogs for 20 epochs (180 exemplars), or on all mammals except the extracted dog, for 2 epochs. (178 exemplars). The result (Fig. 3B) showed that the model adapted significantly faster to the dog when it had only been trained on the other dogs than on all mammals ($F(9) = 3.7973$, $p<.01$). This result provides further evidence that adaptation time (looking time) is affected by experience, and that experience with similar objects reduces adaptation time to novel objects. Conversely, this result also shows that familiarization times are fastest when prior experience is maximally similar to the stimuli tested in the familiarization experiment, and that broader experience can slow down familiarization time compared with narrow experience.

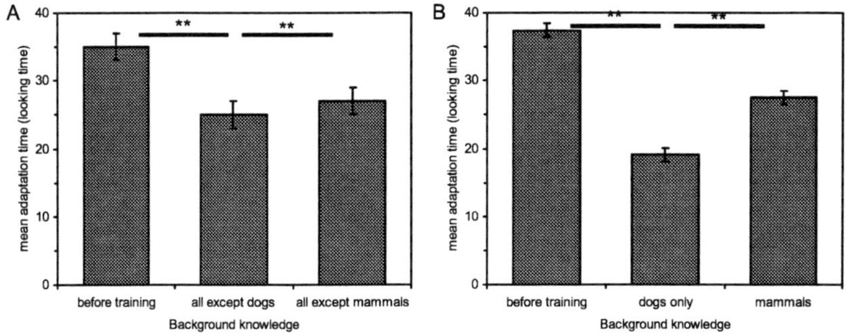

Figure 3: A) Familiarization time to dogs depending on different types of prior knowledge. B) Familiarization time to dogs depending on whether the model had prior experience with dogs. Results are averaged over 10 runs

3. Discussion

The simulations reported in this paper describe preliminary explorations with a multiple memory model of infant category development. However, they raise some important issues about infant studies. Although familiarization/novelty-preference studies examine within-task category formation, the model suggests that presented stimuli in perceptual categorization tasks can activate top-down long-term representations that impact on the formation of representations for the familiarization stimuli. It was not necessary for the models to have background knowledge of the specific basic-level category that was tested in the familiarization task. Instead, experience with other animals was sufficient to speed up familiarization to dogs. These results highlight the need for assessing both the background knowledge of infants and the similarity of experimental stimuli to objects known to the infant.

The model also highlighted that it is important to consider which representations are exploited by different experimental paradigms. Cortical representations in the model show a gradual differentiation from global to basic level categories, suggesting that tasks based on these representations show the same profile.

Together, the model provides initial evidence (see also [23]) that a multiple-memory system perspective provides a useful framework for understanding the development of categorization in infancy.

References

1. Mareschal, D. and P.C. Quinn, *Categorization in infancy.* Trends in Cognitive Sciences, 2001. **4**: p. 443-450.
2. Fantz, R.L., *Visual experience in infants: Decreased attention to familiar patterns relative to novel ones.* Science, 1964. **146**: p. 668-670.
3. Quinn, P.C., P.D. Eimas, and S.L. Rosenkrantz, *Evidence for representations of perceptually similar natural categories by 3-month-old and 4-month-old infants.* Perception, 1993. **22**: p. 463-475.
4. Quinn, P.C. and P.D. Eimas, *Perceptual cues that permit categorical differentiation of animal species by infants.* Journal of Experimental Child Psychology, 1996. **63**: p. 189-211.
5. Mareschal, D., R.M. French, and P.C. Quinn, *A connectionist account of asymmetric category learning in early infancy.* Developmental Psychology, 2000. **36**: p. 635-645.
6. French, R.M., et al., *The role of bottom-up processing in perceptual categorization by 3-to 4-month-old infants: Simulations and data.* Journal of Experimental Psychology-General, 2004. **133**: p. 382-397.
7. Quinn, P.C. and P.D. Eimas, *Perceptual Organization and Categorization in Young Infants*, in *Advances in Infancy Research*, Rovee-Collier and L. C.Lipsitt, Editors. 1996. p. 1-36.
8. Behl-Chadha, G., *Basic-level and superordinate-like categorical representations in early infancy.* Cognition, 1996. **60**: p. 105-141.
9. Quinn, P.C. and M.H. Johnson, *Global-Before-Basic Object Categorization in Connectionist Networks and 2-Month-Old Infants.* Infancy, 2000. **1**: p. 31-46.
10. Mandler, J.M. and L. McDonough, *Drinking and driving don't mix: Inductive generalization in infancy.* Cognition, 1996. **59**: p. 307-335.
11. Mandler, J.M., *Perceptual and Conceptual Processes in Infancy.* Journal of Cognition and Development, 2000. **1**: p. 3-36.
12. Mandler, J.M. and L. McDonough, *On developing a knowledge base in infancy.* Developmental Psychology, 1998. **34**: p. 1274-1288.
13. Quinn, P.C., *Multiple sources of information and their integration, not dissociation, as an organizing framework for understanding infant concept formation.* Developmental Science, 2004. **7**: p. 511-513.

14. McClelland, J.L., B.L. McNaughton, and R.C. O'Reilly, *Why there are complementary learning systems in the hippocampus and neocortex: Insights from the successes and failures of connectionist models of learning and memory.* Psychological Review, 1995. **102**: p. 419-457.
15. Ashby, F.G. and S.W. Ell, *The neurobiology of human category learning.* Trends in Cognitive Sciences, 2001. **5**: p. 204-210.
16. Nelson, C.A., *The ontogeny of human memory - a cognitive neuroscience perspective.* Developmental Psychology, 1995. **31**: p. 723-738.
17. Mareschal, D. and R. French, *Mechanisms of categorization in infancy.* Infancy, 2000. **1**: p. 59-76.
18. Westermann, G. and D. Mareschal, *From parts to wholes: Mechanisms of development in infant visual object processing.* Infancy, 2004. **5**: p. 131-151.
19. Sokolov, E.N., *Perception and the Conditioned Reflex.* 1963, Hillsdale, NJ: Erlbaum.
20. Pauen, S. and B. Träuble. *Does experience with real-life animals influence performance in an objectexamination task?* Presentation given at the Biennial Meeting of the International Conference on Infant Studies; 2004, Chicago, IL
21. Quinn, P.C. and P.D. Eimas, *Evidence for a Global Categorical Representation of Humans by Young Infants.* Journal of Experimental Child Psychology, 1998. **69**: p. 151-174.
22. French, R.M., *Catastrophic forgetting in connectionist networks.* Trends in Cognitive Sciences, 1999. **3**: p. 128-135.
23. Mermillod, M., et al. *The importance of long-term memory in infant perceptual categorization.* in *Proceedings of the 25th Annual Conference of the Cognitive Science Society*. 2004: Mahwah, NJ: Erlbaum.

A DUAL-LAYER MODEL OF HIGH-LEVEL PERCEPTION

J. W. HAN*, PETER C. R. LANE, NEIL DAVEY, AND YI SUN

School of Computer Science, University of Hertfordshire, Hatfield, Hertfordshire, AL10 9AB, United Kingdom

** E-mail: j.w.han@herts.ac.uk*

www.herts.ac.uk

The human visual system is capable of identifying objects because it uses several kinds of information: low-level features and high-level schemata are combined to extract data and identify objects. Here, we present a dual-layer model of high-level perception, which learns both basic features and high-level schemata, combining this information to identify objects. We have tested the dual-layer system by training it to detect faces. The hierarchical knowledge structure and manner in which expectations drive the location of features in an image, make the model a viable process model of high-level perception.

Keywords: Computer Vision; Face Detection; Dual-layer Model; Hierarchy; Feature Map.

1. Introduction

The human visual system takes advantage of two important sources of information: low-level features and high-level schemata. The relation between features and schemata can be viewed as a hierarchical knowledge structure. A key element in human vision is how the knowledge structure is used in both top-down (expectation-driven), and bottom-up (feature-driven) manners. Inspired by this fact, we have developed a model of scene understanding, which we call *Cengji*, after the Chinese word for hierarchy.

Scenes are resolved by the human visual system into components. People then learn the relations among these components to form certain knowledge.[1] This knowledge helps humans to understand a scene in two directions, bottom-up and top-down. In a bottom-up approach the individual features found in a scene are first detected and classified. The features are then grouped together to form larger objects, which then are grouped, sometimes in many levels, until a complete hierarchy is formed. In a top-down approach, an overview of the hierarchy is formed. This overview tells the system which features and objects are likely to form the scene. The system can then direct its attention at the targets in different levels, identifying objects by locating the basic features.

The *Cengji* model aims to utilize human-like vision mechanisms to improve the performance of a computer vision system. It is a novel, trainable computer vision system which can integrate salient features from images into complex, hierarchical representation of objects. The core of this model is the knowledge of objects and the hierarchies among these objects. To distinguish objects and their hierarchical relations needs classifiers. These classifiers can be decided by user arbitrarily. The knowledge and hierarchies in the *Cengji* model can be expanded both by training or manually input. The model is designed to be an expandable and open system. Within this paper, we introduce the *Cengji* model and use an application to face detection to demonstrate the importance of an attention-based mechanism, combining information from high-level schemata with low-level features.

2. Background

Human perception relies considerably upon expectations of what may be found in a scene. It has long been recognized that locating information in a visual scene requires a more active seeking of patterns based on familiar patterns, or schemata. For example, in character recognition, an expectation that letters are from a standard alphabet, or form words, can enable correct identification of otherwise ambiguous data.[2,3] Biederman[4] has also shown that objects are recognized more quickly and more accurately when they fall within a familiar schema. Various cognitive models have been developed of these phenomena,[3,5] but none has reached the level of performance required to perform meaningful identification of objects in a complex application.

Computer vision provides an alternative approach to understanding how to make a computer interpret the visual world. A central research task in this area is how to generate a correct and effective representation of the world. A key feature of many complex systems is the hierarchy.[6] We see increasingly how hierarchical representations are being used in computer vision systems. Dillon *et al.*[7] developed *Cite*, a scene understanding and object recognition system that can generate hierarchical descriptions of visually-sensed scenes based on an incrementally-learnt, hierarchical knowledge base. In a further project, Behnke *et al.*[8] proposed a hierarchical neural architecture for image interpretation, which was based on image pyramids and cellular neural networks inspired by the principles of information processing found in the visual cortex.

Face detection is one of the challenging tasks of computer vision.[9] Support Vector Machines (SVMs) offer the chance for face detection applications to deliver high performance. A direct approach, using single-level SVMs to detect faces, achieved remarkable success when Osuna *et al.*[10] applied SVMs to face detection, later extending this application to a real-time system. El-Naqa *et al.*[11] applied SVMs to detect micro-calcifications in mammograms.

Recently, multi-layer SVMs are becoming popular; these multi-layer SVMs typically use a hierarchical structure to represent information. Heisele *et al.*[12] present a dual-layer SVM algorithm. In this algorithm, component-based face classifiers were combined into a second stage to yield a hierarchical SVM classifier. In the first layer, the component classifiers independently detected components of the face. In the second layer, the combination classifier detected a whole face based on the output of the component classifiers. They also compared the performance between component-based (dual-layer) and global (single-layer) approaches.[13] Their experiments showed the potential of dual-layer SVMs in providing invariant performance with adjustments in pose and illumination.

3. *Cengji*: A Dual-Layer System

We applied *Cengji* to perform object identification. *Cengji* works in the following way. First, it finds individual features from images, then uses the lowest level, bottom-up knowledge of the hierarchy to compose the features together to form a feature map, describing the high-level object. Potentially, these objects can form higher level combination. This is a bottom-up scene understanding process. During this process, the model can also use top-down knowledge stored in the hierarchy. That means when a feature or an object is found, the model uses top-down knowledge to figure out what would be other related features or objects, and where and how to find them. This mechanism helps the system uses its hierarchy effectively in guiding the location and identification of features.

3.1. *System Overview*

We have initially applied the *Cengji* system to the detection of faces, and so the treatment in this section will be in terms of faces. However, we expect to apply *Cengji* to alternative domains, by generalizing the features identified in the first layer. Figure 1 shows the basic structure for detecting faces.

The first layer detects features from the target object. It comprises four

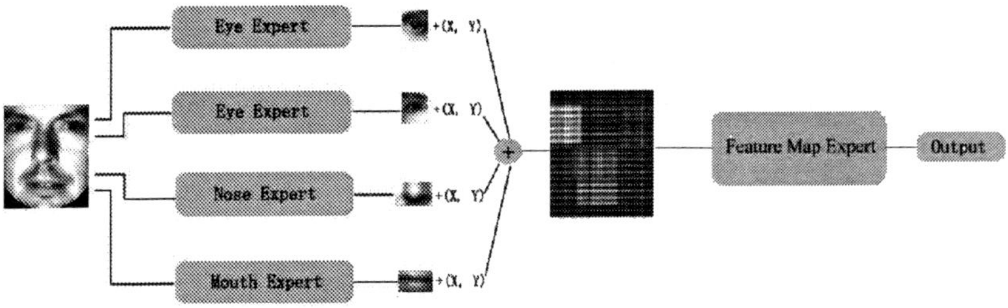

Fig. 1. Structure of *Cengji*

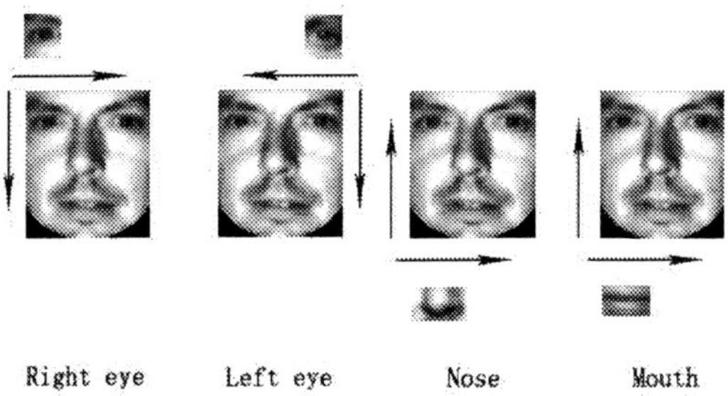

Fig. 2. Grid movement directions.

SVM feature-detection experts: two eyes, nose, and mouth. Two eye experts are trained with the same training sets. This is because the difference between left and right eyes is very small. They are discriminated by different searching sequences in the potential object area. See Figure 2.

By detecting these features, *Cengji* gets feature positions and gives each feature a label: left eye 4, right eye 1, nose 3 and mouth 2. The background is 0. Each number in the feature map is a feature label and corresponds to a pixel. So the map matrix has the same size as the image. There is overlap between two different features. *Cengji* can learn this because the feature detection is always in the same sequence: right eye → mouth → nose → left eye. This ensures that the overlap between two features is consistent. Figure 3 is a feature map for a face.

The second layer, also an SVM classifier, is used to identify the structure of how these features compose the target object. Its input is the feature map and the output is a class index: 1 for face, 0 for non face. *feature map* acts as a medium to convey object structure directly; it has the advantage of

```
00000000    Key:
01100440
01100440    0 is no feature
00000000    1 for right eye
00003000    2 for mouth
00003000    3 for nose
00000000    4 for left eye
00222200
```

Fig. 3. Simplified feature map for a face; actual size is 84x96 pixels.

preserving all relative positions between features explicitly.

The SVM package we are using is *LibSVM*, which is developed by Chang and Lin.[14]

3.2. *Attention Mechanism*

Attention is the cognitive process of selectively concentrating on one aspect of the environment while ignoring other things [cited from en.wikipedia.org]. This mechanism allows human to concentrate on certain task without being disturbed by useless information. This process also plays a role in machine learning systems, to avoid confusion from irrelevant features.[15]

Most of current computer vision systems treat all kinds of information in object area with the same emphasis. This situation spends limited computation resource on some useless calculation. *Cengji* model only pays attention to certain features relative to the target object without spending more time to useless information. This mechanism is achieved by the utilization of high-level schemata. Figure 4 is a flow chart of the attention mechanism.

The high-level schemata are stored in an associative memory. An autoassociator is trained to store the feature maps associated with faces, and is used to predict the location of features, in the following way. Once the first layer identifies one of the features, say the left eye, the partial feature map will be passed to the autoassociator. The partial feature map will be 'completed', with the associative memory retrieving an example feature map comprising all the related features making up a face. This completed feature map is then used to predict the location of the other features. The *Cengji* system may then guide the first layer experts directly to a hopefully relevant part of the image, without having to do the scanning described above. In this way, *Cengji* can speed up its search for features. *Cengji* will

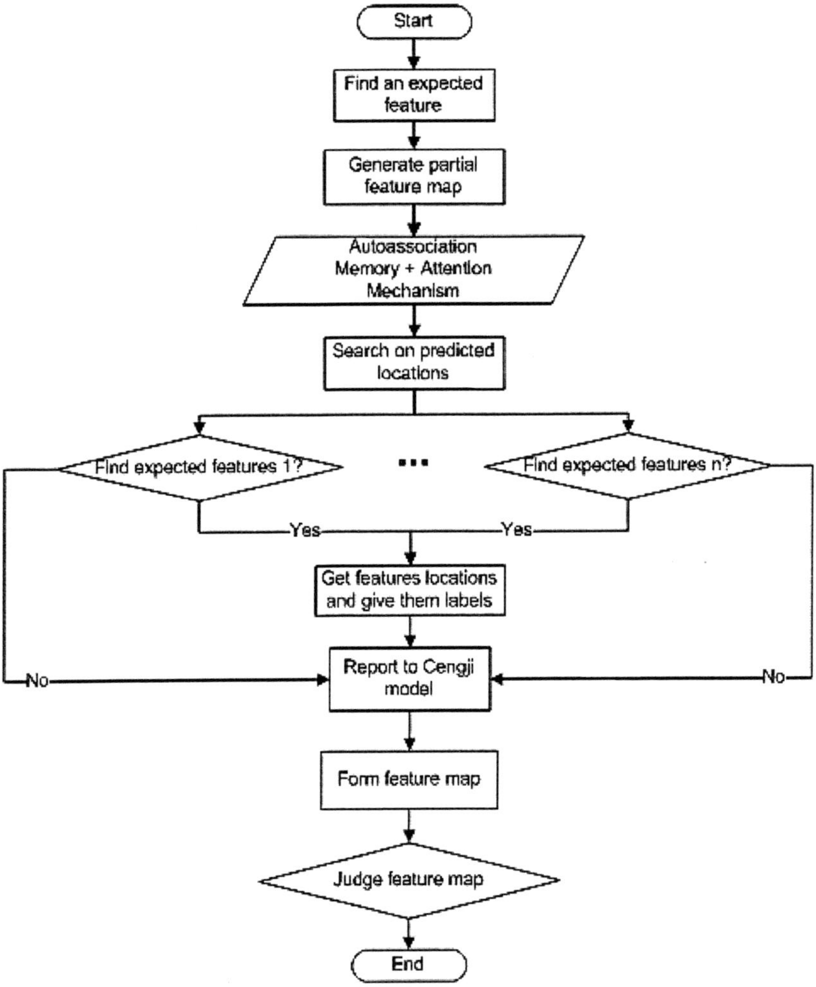

Fig. 4. Process of Attention Mechanism.

also avoid a problem with false positives, where features are identified which are not in a position relevant to the first feature in order to make up a face. Figure 5 is a sample of detecting different features from face. The right side is a detected face in which only the first detected features were taken to form the face.

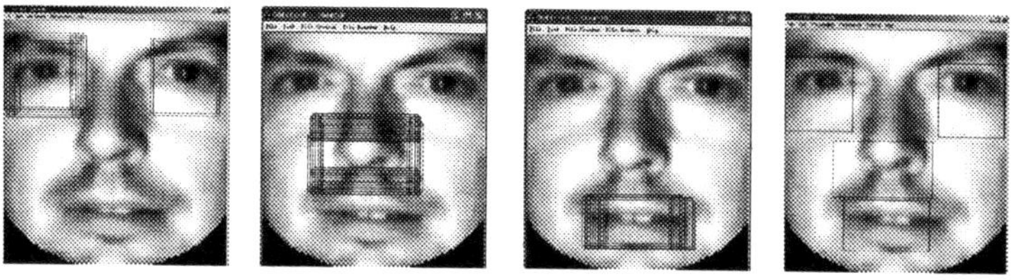

Fig. 5. Detection of features

4. Databases

In this section, we describe the databases used in the following experiments, and the pre-processing we had to perform to construct datasets for each of the feature experts.

We used four following databases to create positive training samples: Database of Faces from AT&T Laboratories, Cambridge; Japanese Female Facial Expression (JAFFE) Database; and The Psychological Image Collection at Stirling (PICS). The following databases were used to create negative training samples: BEV1 Dataset and Caltech Database.

We cut faces from these datasets and adjusted the face image size to a consistent size of 84×96 pixels. The proportion of faces to the image size is consistent. We randomly selected 201 face images for training, 178 for test, and 451 non-face images for training, 227 for test. Especially, we selected 78 frontal faces in test set, which have attitude extremity no more than 15^0 in all directions.

For the training of feature experts, we located the features image by image manually from the training images described above. Feature sizes were decided by their geometry characteristics. For the eye, every eyebrow and eye from all images should be contained within the feature area. So 25×29 pixels is the size which contains the biggest eyebrow and eye pair. The size of nose, 38×22 pixels, is decided by the nose width and the distance from bridge of a nose to the very bottom of nose. The width and the height of mouth decide the size, 33×20 pixels, of mouth samples.

Creating a negative dataset is challenging, because it is difficult to get key negative images, which are close to positive images; learning is more efficient when the negative images are 'close' to the positive images. We used an iterative process to create a negative dataset, allowing the learning algorithm itself to locate the 'near-misses':

(1) Select some non-feature images, which form the basic negative samples. These have a similar average grey scale to the faces in the training set.

Table 1. Feature Datasets details

Feature	Number of Images		Image size
	Positive	Negative	
Eye	344	1865	25 × 29
Nose	203	1250	38 × 22
Mouth	201	1634	33 × 20

(2) Select some negative examples from the feature images to act as negative samples for other features. For example, noses and mouths as negative samples for eyes.
(3) Use the dataset to train *Cengji*. Then use the trained system to recognize features from face dataset. Find the falsely detected features and add them to the negative dataset. At the same time, add those not detected features to positive training set. We may delete some feature images or repeatedly cut features from one face image.
(4) Repeat step 3 until performance acceptable on training set.

Table 1 shows details of these training sets.
Figures 6 shows examples in eye training set.

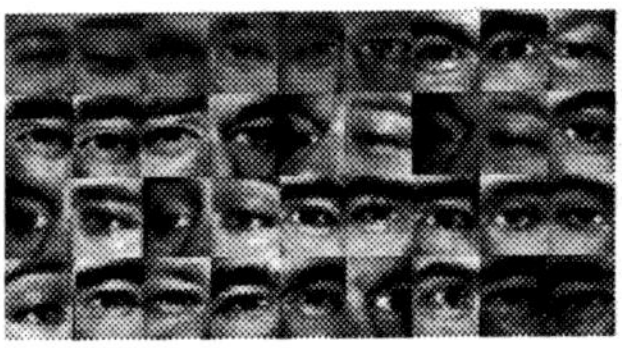

Fig. 6. Part of eye datasets. Left: positive eye examples. Right: negative eye examples.

For the second layer, positive and negative feature maps were created artificially by giving each feature different situations in the feature map. In order to create positive dataset, different tolerance was given to each feature which allowed each feature to change its situation in a certain area. The feature map was treated as a negative sample if some features were outside of its positive area.

5. Experiments

The aim of these experiments is to compare the performance of the Cengji system, with and without its attention mechanism. First, we confirm that the feature experts in the first layer work effectively. Second, we perform experiments on the complete system.

Table 2. Test results of each feature expert

	Eyes	Noses	Mouths
Correct(%)	94.08	96.88	92.12

5.1. *Performance of the feature experts*

The first layer of *Cengji* comprises 4 independent feature experts. The performance of the feature experts is important, as they form the data on which the feature map, and the system's top-level classifications and knowledge, is created.

Method. We tested the performance of each feature expert individually. The *LibSVM* library offers a range of parameters for optimizing the performance of each expert. We performed a grid search across the parameter space, and found that the RBF kernel performed the best with the remaining parameters at their default settings.

For each expert, we kept 25% of the total dataset as a held-out test set. The remaining 75% was used to do 5-fold cross validation. This gives a statistical evaluation on the performance of each expert. In the training and test set, the positive and negative numbers are not balanced. We weighted the trade-off coefficient C for the class with less samples to reduce this imbalance; this prevents any bias of the learning system towards the more popular class.

Results. Table 2 shows test results on the held-out test set. Note that all three feature experts produce good results, over 90%.

Discussion. The feature experts play an important role in classifying complex images. The test results here indicate that the individual feature experts perform well on identifying examples of their features. However, when scanning a complete image, such as a face, there are many more parts of the face which are not eyes than there are parts which are eyes. Hence, the small error rate for the eye feature detector can translate into many falsely detected eyes in a complete image. We will see the effect of this in the next experiment.

5.2. *Comparing the Performance of* Cengji *with and without Attention Mechanism*

We evaluated the performance of the complete *Cengji* system, both with and without the attention mechanism. The aim is to explore the impact of an attention mechanism on classification of visual images.

Method. First, we compared two models on test set containing 78 frontal

Table 3. Confusion matrix of *Cengji*, with and without AM. The row label gives the correct class, and the column label the class produced by the system.

	With AM		Without AM	
	Face	Non-face	Face	Non-face
Face	71	7	63	15
Non-face	15	63	16	62

faces and 78 non-faces. Then we used method introduced in Ref. 16 Chapter 5. In this comparison, we used the training set only. Each feature set was partitioned into 5 subsets. Each time, one subset of each feature was picked out and faces where features in this subset were from were taken as test samples. Features picked out were all from same faces. The remaining subsets were used to train *Cengji*. This process was repeated until every subset was tested. Each time we calculated the error difference generated by two classifiers and at the end got the average error difference.

Results. Table 3 is the confusion matrix. It shows the *Cengji* with autoassociation memory (AM) gave a better performance on frontal faces: 91.03% correct. Without autoassociation memory, the correct rate was 80.77%. On the non-faces, they were equivalent: 80.77% vs. 79.49%.

During the 5-fold cross test, the average error difference between them was 1.02%. *Cengji* with AM got an average error rate 19.35%. Without AM, it was 20.37%. This means, without AM, the performance of *Cengji* was only slightly worse on detecting both frontal and side faces. The reason is there are some side faces but the *Cengji* only has association memory on frontal faces. The frontal face association memory was not working properly on side faces. This indicates the effect of AM from another aspect.

Discussion. The results here indicate that the main effect of adding the attention mechanism is to convert many of the false negatives into true positives. In other words, the attention mechanism enables *Cengji* to avoid misclassifying certain faces as non-faces. Figure 7 gives an example of the face correctly classified with the attention mechanism.

Following on from the discussion for the previous experiment, we see that the attention mechanism reduces the detection of, for example, non-noses as noses. This means that the feature map is more likely to have features within the correct relative positions, which the second level can correctly identify as a face. A similar effect is not found for the non-faces.

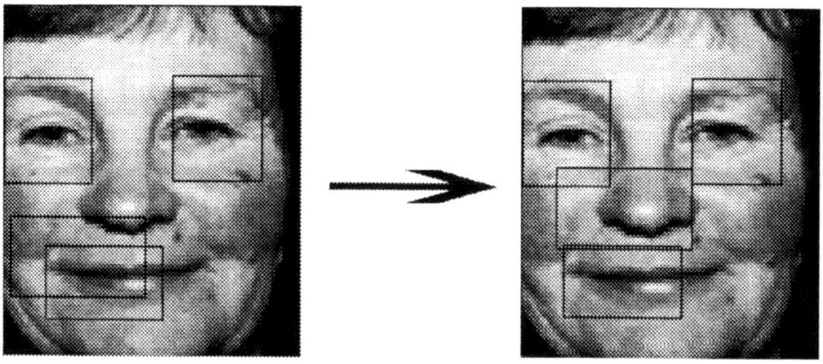

Fig. 7. Comparison. Left is a result got by *Cengji* without AM. The nose location is wrong. Right, *Cengji* with AM correctly located the nose detected it as a face.

5.3. *Discussion*

The attention mechanism provides *Cengji* with a form of top-down, schematic knowledge of the objects which it is attempting to recognize. This top-down knowledge is important in helping to prevent the misclassification of known objects because features are wrongly identified in unexpected positions. We saw an example of this in Figure 7.

The *Cengji* system, with its attention mechanism, fits with psychological theories of the role of expectations in human perception. We discussed earlier how one of the characteristics was that objects are recognized more accurately when they fall within a familiar schema;[4] this characteristic has been found to be true of *Cengji* in our experiments.

6. Conclusion and Future Work

Component-based models for machine vision are proving increasingly popular. In this paper we have taken inspiration from the psychology of human perception to include the role of schemata in guiding the low-level identification of features within an image. We have introduced a feature map and an associative memory for the schemata. Experimental results show that the basic processes are sound, producing good results in a simple task of identifying faces. Future work will explore several avenues, the most important of which is to include more kinds of object. One limitation of the system at present is that the initial, feature level needs to be designed. We anticipate that feature-construction techniques may enable the system to locate its own useful features and so learn about different kinds of objects. A further interesting experiment would be to see how well *Cengji*'s direction of feature locations matches that of humans.

References

1. I. Biederman, *Psychological Review* **94**, 115 (1987).
2. U. Neisser, *Cognitive Psychology* (New York: Appleton-Century-Crofts, 1966).
3. H. B. Richman and H. A. Simon, *Psychological Review* **3**, 417 (1989).
4. I. Biederman, On the semantics of a glance at a scene, in *Perceptual Organization*, eds. M. Kobovy and J. R. Pomerantz (Hillsdale, NJ: Lawrence Erlbaum, 1981) pp. 213–254.
5. P. C. R. Lane, A. K. Sykes and F. Gobet, Combining low-level perception with expectations in CHREST, in *Proceedings of EuroCogsci*, eds. F. Schmalhofer, R. M. Young and G. Katz (Mahwah, NJ: Lawrence Erlbaum Associates, 2003).
6. H. A. Simon, *The sciences of the artificial* (Cambridge (Mass.) ; London : M.I.T. Press, 1969).
7. C. Dillon and T. Caelli, *Journal of Computer Vision Research* **1**, 89 (1998).
8. S. Behnke and R. Rojas, Neural abstraction pyramid: a hierarchical image understandingarchitecture, in *The 1998 IEEE International Joint Conference on Neural Networks Proceedings. IEEE World Congress on Computational Intelligence.*, 1998 pp. 820–825.
9. M. H. Yang, D. J. Kriegman and N. Ahuja, Detecting faces in images: A survey, in *IEEE Transactions on Pattern Analysis and Machine Intelligence*, 2002 pp. 34–58.
10. M. A. Hearst, *IEEE Intelligent Systems* **13**, 18 (1998).
11. I. El-Naqa, Y. Y. Yang, M. N. Wernick, N. P. Galatsanos and R. Nishikawa, Support vector machine learning for detection of microcalcifications in mammograms, in *IEEE Transactions on Medical Imaging*, 2002 pp. 1552–1563.
12. B. Heisele, T. Serre, M. Pontil, T. Vetter and T. Poggio, Categorization by learning and combining object parts, in *Advances in Neural Information Processing Systems*, 2001 pp. 1239–1245.
13. B. Heisele, P. Ho, J. Wu and T. Poggio, *Computer Vision and Image Understanding* **91**, 6 (2003).
14. C. C. Chang and C. J. Lin, Libsvm – a library for support vector machines http://www.csie.ntu.edu.tw/ cjlin/libsvm/.
15. A. L. Blum and P. Langley, *Artificial Intelligence* **97**, 245 (1997).
16. T. M. Mitchell, *Machine Learning* (The McGraw-Hill Companies, Inc., 1997).

Section IV

Sensory and Attentional Processing

PROCESSING SYMBOLIC SEQUENCES USING ECHO-STATE NETWORKS

MICHAL ČERŇANSKÝ
Faculty of Informatics and Information Technologies, Slovak University of Technology, Ilkovičova 3, 842 16 Bratislava 4, Slovakia

PETER TIŇO
School of Computer Science, University of Birmingham, Edgbaston, Birmingham B15 2TT, United Kingdom

A novel recurrent neural network (RNN) model, called the echo state network (ESN), was successfully applied in several time series processing tasks. Also using ESNs for processing symbolic sequences seems to be a promising possibility. But ESNs, like RNNs initialized with small weights, share some properties with variable length Markov models when used for processing symbolic sequences. We call this phenomenon a Markovian architectural bias since meaningful and potentially a useful Markov-like state space organization is present in an RNN prior to any training. In this paper we first explain this notion in detail. Then, in an experimental section, we compare the performance of ESNs with connectionist models explicitly using the Markovian architectural bias property and with variable length Markov models (VLMMs). We show that ESNs performance remains almost the same for a wide range of parameters and that the number of reservoir units plays a similar role to the number of context units of other models. We show that ESNs, like other connectionist models with state space organized according to the Markovian architectural bias property, cannot generalize well on subsequences not present in the training set.

1. Introduction

A lot of attention is now being focused on connectionist models known under the name "reservoir computing". The most prominent example of these approaches is a recurrent neural network (RNN) architecture called echo state network (ESN). The recurrent layer of an ESN consists of a large number of sparsely interconnected units with non-trainable weights and serves as the reservoir of potentially interesting behavior. Computationally less demanding adaptation processes are required since only output weights are adjusted to fit the training data. ESNs were successfully applied in many real-valued time series modeling tasks and performed exceptionally well.

Much commonly used real-world data with time structure can be expressed as a sequence of symbols from a finite alphabet, i.e., a symbolic time series. Since their emergence, neural networks were applied to symbolic time series analysis. Especially popular is to use connectionist models for processing of complex language structures. Some attempts were made to process symbolic time series using ESNs with interesting results. ESNs were trained on stochastic symbolic sequences and a short English text in [1] and were compared with other approaches including Elman's simple recurrent network (SRN) trained by a simple BP algorithm in [2]. Also other researchers have expressed their interest in applying ESNs for processing symbolic sequences without realizing that the specific organization of the network's state space prevents the network from generalizing on unseen patterns.

The article is organized as follows: In Sec. 2 we introduce ESNs as a special type of RNNs. Sec. 3 is devoted to the Markovian architectural bias of RNNs. Here we explain the organization of the state space of RNNs initialized with small weights, when they are used for processing symbolic time series. In Sec. 4 we present results of experiments done with two symbolic time series. Predictive performance of ESNs is compared with other models such as VLMMs and prediction machines directly using Markovian architectural bias property.

2. Echo State Networks

RNNs were successfully applied in many real-life applications where processing time-dependent information was necessary. Unlike feedforward neural networks, units in RNNs are fed by activities from previous time steps through recurrent connections. In this way contextual information can be kept in units' activities, enabling RNNs to process time series.

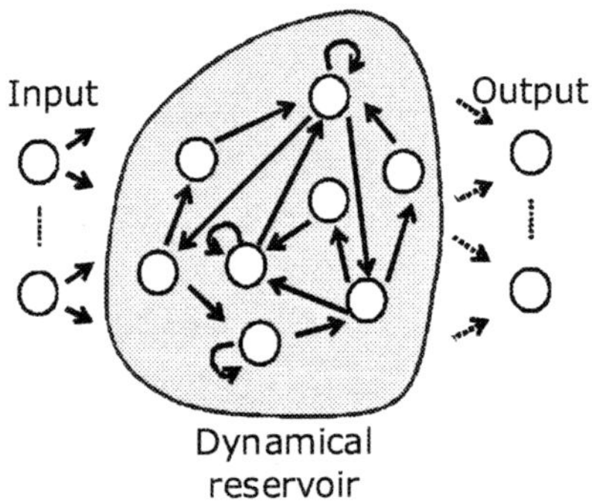

Figure 1. Echo state network architecture. Dashed arrows indicate trainable weigths.

Echo state networks represent a new powerful approach in recurrent neural network research [3,4]. Instead of a difficult learning process, ESNs are based on the property of untrained randomly initialized RNNs to reflect the history of seen inputs - here referred to as "echo state property". An ESN can be considered to be an SRN with a large and sparsely interconnected recurrent layer – "reservoir" of complex dynamics. Only the network's output connections are modified during learning process in order to produce desired responses by extracting interesting features from ESN dynamics. A significant advantage of this approach is that computationally effective linear regression algorithms can be used for adjusting the output weights.

The network includes input, hidden and output "classical" sigmoid units as shown in Fig. 1. The reservoir of the ESN dynamics is represented by hidden layer with partially connected hidden units. The main and essential condition for successful use of the ESN is the "echo state" property. The network state is required to "echo" the input history. If this condition is met, only adaptation of network output weights is required to obtain an RNN with high performance. However, for large and rich reservoir of dynamics, hundreds of hidden units are needed.

When x is an input vector at time step t, activations of internal units are updated according to

$$x(t)=f\left(W^{in}\cdot u(t)+W\cdot x(t-1)+W^{back}\cdot y(t-1)\right), \tag{1.1}$$

where f is the internal unit's activation function, W, W^{in} and W^{back} are hidden-hidden, input-hidden, and output-hidden connections' matrices, respectively. Activations of output units are calculated as follows:

$$y(t)=f\left(W^{out}\cdot\left[u(t),x(t),y(t-1)\right]\right), \tag{1.2}$$

where W^{out} is output connections' matrix.

Echo state property means that for each internal unit x_i there exists an echo function e_i such that the current state can be written as $x_i=e_i\left(u(t),u(t-1),\ldots\right)$ [4]. The recent input presented to the network has more influence to the network state than an older input, the input influence gradually fades out. So the same input signal history $u(t),u(t-1),\ldots$ will drive the network to the same state $x(t)$ in time t regardless the network initial state.

3. Markovian Architectural Bias of Recurrent Neural Networks

3.1. *Organization of the RNN State Space Initialized with Small Weights*

The dynamics of RNNs initialized with small weighs shares some properties with Markov models [10]. Imagine simplified RNN dynamics obtained by using linear activation function in Eq. (1.1) and setting recurrent and input weight matrices to:

$$W = \begin{pmatrix} 0.5 & 0 \\ 0 & 0.5 \end{pmatrix}, \qquad W^{in} = \begin{pmatrix} 0.5 & 0 & 0 & 0.5 \\ 0 & 0.5 & 0 & 0.5 \end{pmatrix} \tag{1.3}$$

No backwards weights were used. Processing sequence of symbols $s(1), s(2)\ldots s(t)$ created over 4 symbol alphabet $A = \{a,b,c,d\}$ with symbols encoded by one-hot-encoding scheme (if $s(t)=a$ than $u(t)=(1,0,0,0)^T$) would result into the partitioning of the state space as shown in Fig. 2.

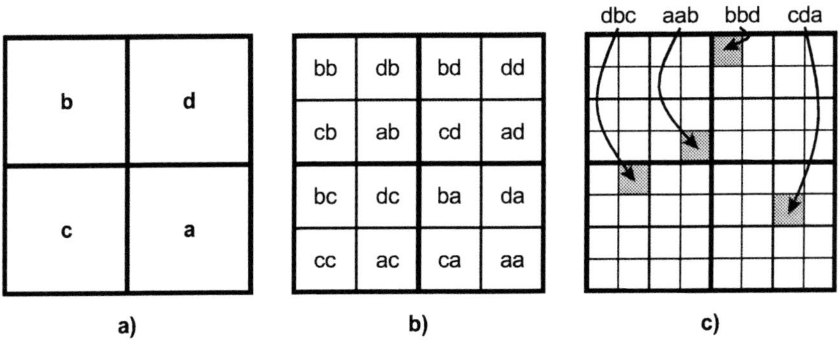

Figure 2. The relationship between the network state and the history of symbols presented to the network with simplified dynamics.

The network state (activities of hidden units) is always located in a region corresponding to the last symbol $s(t)$ presented to the network as shown in Fig. 2a. Moreover the position within this region can be further specified by the previously presented symbol $s(t-1)$. This "second level" is shown in Fig. 2b and sub-regions differentiated by the symbol presented to the network in time $s(t-2)$ are shown in Fig. 2c. The position of activities in the network state space is determined by the history of symbols presented to the network with the recent symbols having more important impact than symbols presented in older time steps.

To obtain a visual representation of the whole sequence we can insert hidden unit activities $s(t)$ from all time steps into two-dimensional plot as shown in

Fig. 3. This way of representing input sequence is called a chaos game representation (CGR). The random sequence was created by drawing symbols from alphabet A with equal probability. A laser data set is the sequence of differences between successive activations of a real laser in chaotic regime [5] which was quantized to symbolic sequence over four-symbol alphabet A. It is described in more details in Sec. 4.1. CGR of random sequence results in the state space regularly covered by points. It is not the case of CGR of the laser sequence. Similar subsequences correspond to points that are closer in the state space. The longer the common suffix, the nearer the points are in the state space. Frequent subsequences of longer length produce clusters.

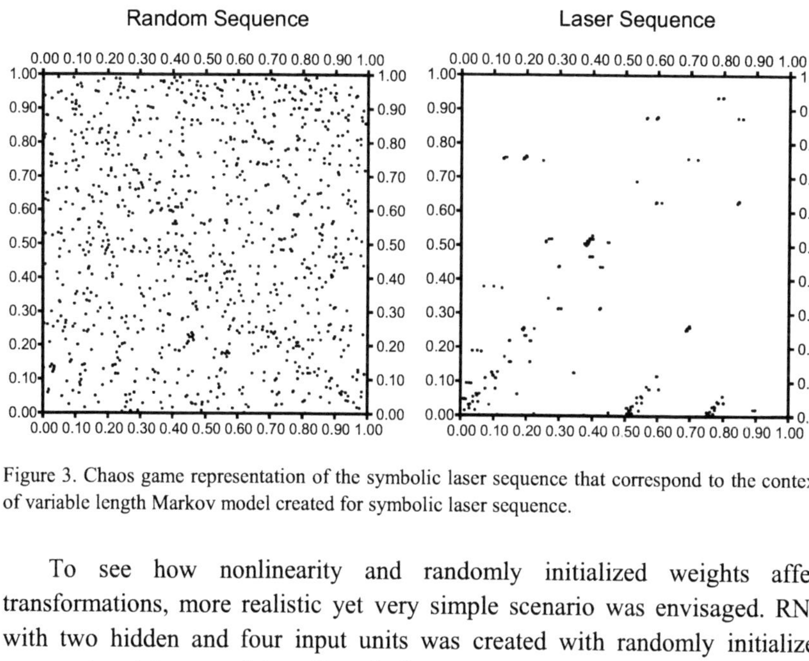

Figure 3. Chaos game representation of the symbolic laser sequence that correspond to the contexts of variable length Markov model created for symbolic laser sequence.

To see how nonlinearity and randomly initialized weights affect transformations, more realistic yet very simple scenario was envisaged. RNN with two hidden and four input units was created with randomly initialized recurrent and input weights. Hyperbolic tangent activation function was used. Transformations of the whole state space $\langle -1,1 \rangle^2$ driven by input symbols are shown in Fig. 4. We can clearly see that the state space is partitioned according the four symbols presented to the network in the same way as show in Fig. 2.

3.2. *Models Using Markovian Architectural Bias Property*

This organization of the RNN state space has led to the idea described in [5] where prediction models called neural prediction machine (NPM) and fractal prediction machine (FPM) were suggested. Both use Markovian dynamics of untrained recurrent network. In FPM, activation function f is linear and weights are defined deterministically in order to create precise state space dynamics similar to Eq. (1.3). In NPM, activation function f is nonlinear and weights are randomly initialized to small values as in regular RNN. Instead of using classical output layer NPM and FPM use prediction model that is created by extracting clusters from the network state space. Each cluster corresponds to different prediction context with the next symbol probabilities associated to it.

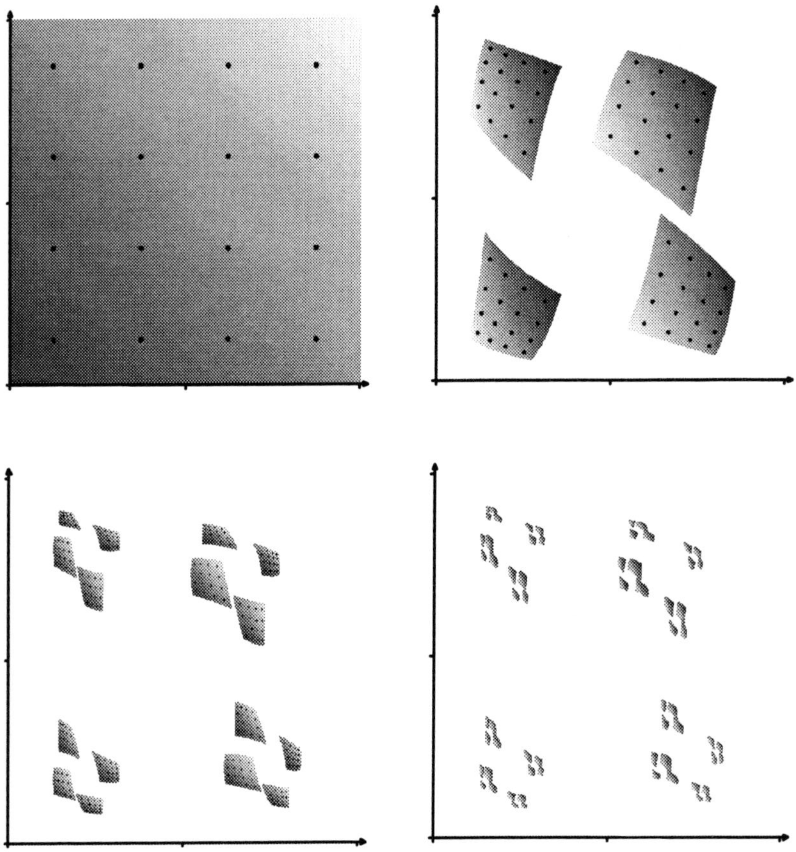

Figure 4. The relationship between the network state and the history of symbols presented to the network for "real" RNN with tanh activation function and randomly initialized small weights.

More precisely, symbol presented to the connectionist model drives it to some state (activities on hidden units). The state belongs to some cluster and the context corresponding to this cluster is used for the prediction. The context's next symbol probabilities are estimated during training process by relating the number of times that the corresponding cluster C is encountered and the given next symbol x is observed (N_C^x) with the total number when the cluster C is encountered (N_C):

$$P(x \mid C) \approx \frac{N_C^x}{N_C}. \tag{1.4}$$

3.3. *Variable Length Markov Models*

Described models share some properties with variable length Markov models. Building Markov next-symbol prediction model is straightforward by estimating probabilities $P(x \mid w) \approx N_w^x / N_w$, where $P(x \mid w)$ is the probability of symbol $x \in A$ being observed after sequence $w = s_1 s_2 \ldots \in A^*$. In this case sequence of observed symbols w represents the prediction context. In n-th order Markov model the set of contexts is formed of all words of length n which can be created from alphabet A. The set contains $|A|^n$ contexts where $|A|$ is the size of the alphabet.

Variable length Markov models (VLMMs) [6,7] contain contexts with different lengths. They try to evaluate the context importance and irrelevant contexts are not kept by the model. The importance of the context is estimated from the training sequence. There is a natural idea of growing in VLMMs. Existing context $w \in A^*$ is extended by symbol $x \in A$ and new context xw is added to the model if its next symbol probability distribution $P(. \mid xw)$ significantly differs from the next symbol probability distribution of original context $P(. \mid w)$ and its occurrence $P(xw)$ is important enough. A natural way of representing VLMMs is in the form of a prediction suffix trees (PSTs). Fig. 5 shows simple PST created for laser sequence. For example the probability of observing symbol b after the sequence $ccca$ will be estimated as $P(b \mid ccca) = 0.06$.

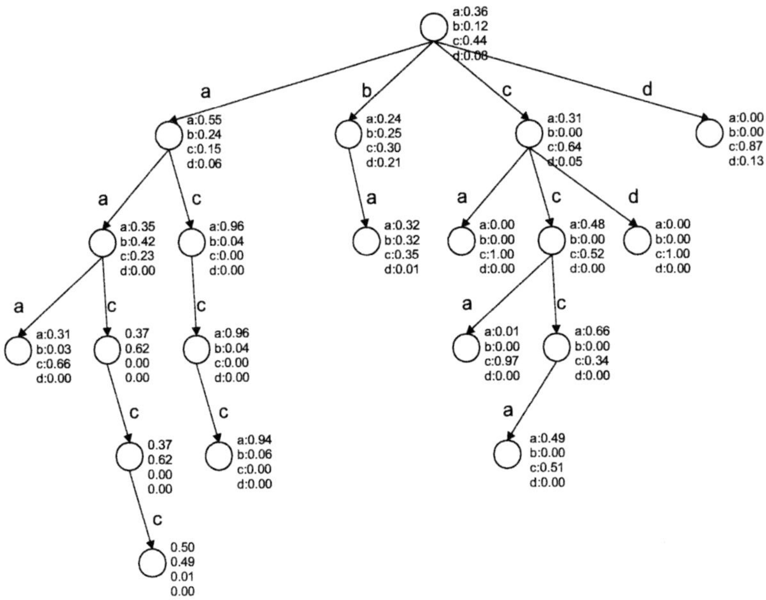

Figure 5. Simple variable length Markov model created for the laser sequence.

4. Experiments

4.1. *Method and Datasets*

We present experiments with two symbolic sequences. The first one was created by symbolization of activations of laser in chaotic regime. The second sequence is inspired by the research in cognitive science and contains words generated by simple context free grammar.

The laser dataset was obtained by quantizing activity changes of laser in chaotic regime, where relatively predictable subsequences are followed by hardly predictable events. The original real-valued time series was composed of 10000 differences between the successive activations of a real laser. The series was quantized into a symbolic sequence over four symbols corresponding to low and high positive/negative laser activity change. The first 8000 symbols are used as the training set and the remaining 2000 symbols form the test data set [8].

Deep Recursion data set is composed of strings of context-free language L_G. Its generating grammar is $G = \{R, \{a, b, A, B\}, P, R\}$, where R is the single non-terminal symbol that is also the starting symbol, and $\{a, b, A, B\}$ is a set of terminal symbols. The set of production rules P is composed of three simple rules: $R \rightarrow aRb \mid ARB \mid e$ where e is the empty string. This language is in [9] called palindrome language. The training and testing data sets consist of 1000 randomly generated concatenated strings. No end-of-string symbol was used. Shorter strings were more frequent in the training set than the longer ones. The total length of the training set was 6156 symbols and the length of the testing set was 6190 symbols.

Predictive performance was evaluated by means of a normalized negative log-likelihood (NNL) calculated over the test symbol sequence $S = s_1 s_2 \ldots s_T$ from time step $t = 1$ to T as

$$\mathrm{NNL} = -\frac{1}{T}\sum_{t=1}^{T} \log_{|A|} p(t), \tag{1.5}$$

where the base of the logarithm is the alphabet size, and the $p(t)$ is the probability of predicting symbol $s(t)$ in the time step t.

In case of ESNs for NNL error calculation the activities on output units were first adjusted to chosen minimal activity $o_{\min}$ set to 0.001 in this experiment, then the output probability $p(t)$ for NNL calculation was evaluated as:

$$\hat{o}_i(t) = \begin{cases} o_{\min} & \text{if } o_i(t) < o_{\min} \\ o_i(t) & \text{otherwise} \end{cases}, \qquad p(t) = \frac{\hat{o}_i(t)}{\sum_j \hat{o}_j(t)}, \tag{1.6}$$

where $o_i(t)$ is the activity of the output unit i in time t.

4.2. *Reservoir of Echo State Networks*

ESNs with hidden unit count varying from 1 to 1000 were trained using recursive least squares algorithm. Symbols were encoded using one-hot-encoding, i.e. all input or target activities were set to 0, except the one corresponding to given symbol, which was set to 1. Hidden units had sigmoidal activation function and linear activation function was used for output units. Reservoir weight matrix was rescaled to different values of spectral radius from 0.01 to 5. The probability of creating input and threshold connections was set to 1.0 in all experiments and input weights were initialized from interval (-0.5,0.5). Probability of creating recurrent weights was 1.0 for smaller reservoirs and 0.01 for larger reservoirs. It was found that this parameter has very small influence to the ESN performance (but significantly affects simulation time).

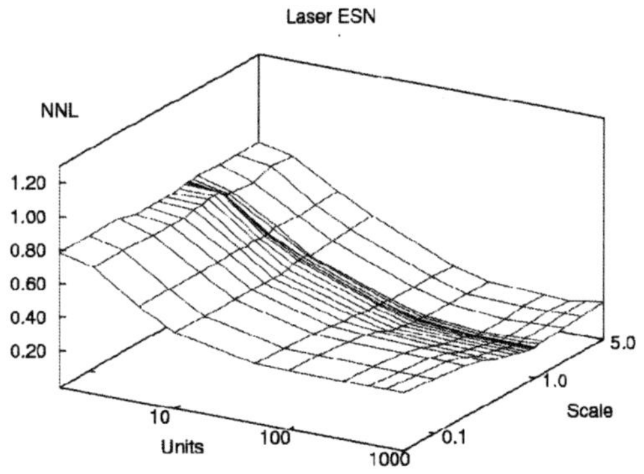

Figure 6. Performance of ESNs with different unit counts and different values of spectral radius.

Results for laser datasets are shown in Fig. 6. Results are very similar for wide range of spectral radii. Having more units in the reservoir results in better prediction. Exactly the same behavior was observed for deep recursion dataset.

For wide range of combinations of reservoir parameters very similar results were obtained. This observation is in accordance with the principles of Markovian architectural bias. Fractal organization of the recurrent neural network state space is scale free and as long as the state space dynamics remains contractive the clusters reflecting the history of the symbols presented to the network are still present.

4.3. *Comparison of ESNs, VLMMS and Models Explicitly Using Markovian Architectural Bias Property*

Fractal prediction machines were trained for the next symbol prediction task on the two datasets. Also neural prediction machines built over the untrained SRN and the trained SRN were used. Prediction contexts for all prediction machines (FPMs and NPMs) were identified using K-means clustering with cluster count varying from 1 to 1000. 10 simulation were performed and mean and standard deviation are shown in plots. Neural prediction machines uses dynamics of randomly created networks. For fractal prediction machines internal dynamics is deterministic. Initial clusters are set randomly by K-menas clustering hence slightly different results are obtained for each simulation also for FPMs. VLMMs were constructed with the number of context smoothly varying from 1 context to 1000 contexts. Details on training SRN can be found in [10]. Results are shown in Fig. 7 and 8.

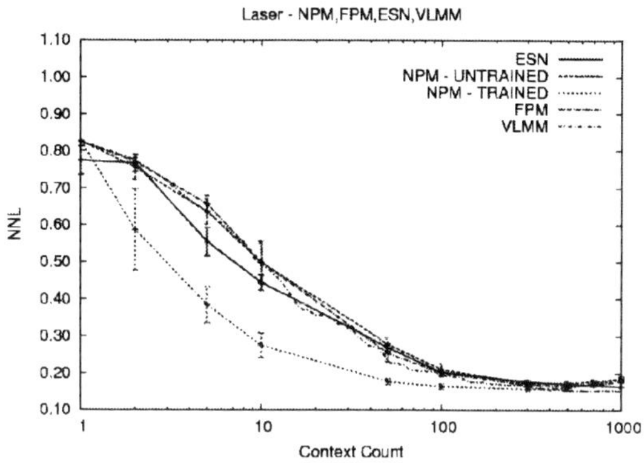

Figure 7. Performance of ESNs compared to FPMs, NPMs and VLMMs for laser dataset.

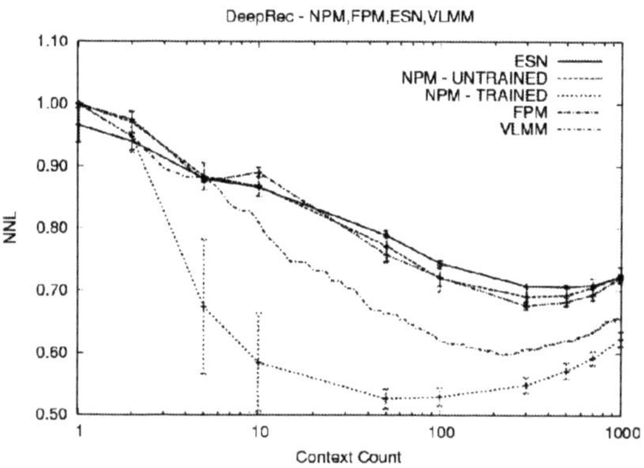

Figure 8. Performance of ESNs compared to FPMs, NPMs and VLMMs for deep recursion dataset.

The first observation is that the ESNs have the same performance as other models using architectural bias properties and that the number of hidden units plays very similar role as the number of contexts of FPMs and NPMs built over untrained SRN. For laser dataset incrementing the number of units resulted in prediction improvement. For deep recursion dataset and higher units count (unit counts above 300) ESN model is over-trained exactly as other models. ESN uses linear readout mechanism and the more dimensional state space we have the better hyper-plane can be found with respect to the desired output.

Training can improve the state space organization so better NPM models can be extracted form the recurrent part of the SRN. For laser dataset the improvement is present for models with small number of context. For higher values of context count the performance of the NPMs created over the trained SRN is the same as for other models. But the carefully performed training process using advanced training algorithm significantly improves the performance of NPMs built over the trained SRN for the deep recursion dataset.

5. Conclusion

In several articles ESNs were successfully used for processing real-valued time series and excellent results were achieved on some datasets. Naturally ESNs were and will be used also for processing symbolic time series. But ESNs in the same way as all other models based on Markovian representation of the inputs can't generalize well on patterns not present in the training set. In this paper we experimentally support this claim. We showed that (apart from reservoir size) ESN reservoir parameters if kept in reasonable intervals have very slight influence on ESN performance when processing symbolic time series. We showed correspondence between the number of units in ESN reservoir and the context count of FPMs, NPMs and Markov models. ESNs are not able to beat Markov barrier when processing symbolic time series. Carefully trained RNNs can achieve better results on certain datasets. On the other side computational expensive training process may not be justified on other datasets and models such as ESNs can perform just as well as thoroughly trained RNNs.

Acknowledgments

This work was supported by the grant APVT 20-030204 and VG 1/4053/07.

References

1. H. Jaeger, Tech.Rep., Ger. Nat. Res. Cent. for Inf. Tech. **GMD152**, (2001).
2. S. Frank, Conn. Science **18**, (2006).
3. H. Jaeger, Tech.Rep., Ger. Nat. Res. Cent. for Inf. Tech. **GMD148**, (2001).
4. H. Jaeger and H. Hass, Science, **304**, (2004).
5. P. Tiňo and G. Dorffner, Int. ICSC/IFAC Sym. on Neur. Comp. (1998).
6. D. Ron, Y. Singer and N. Tishbi, Mach. Learn. **25**, (1996).
7. M. Machler and P. Buhlmann, Journ. of Comp. and Graph. Stat. **13**, (2004).
8. P. Tiňo and G. Dorffner, Mach. Learn. **45**, (2001).
9. P. Rodriguez, Neur. Comp. **13**, (2001).
10. P. Tiňo, M. Čerňanský, Ľ. Beňušková, IEEE Tr. On Neur. Net. **15**, (2004).

NEURAL MODELS OF HEAD-DIRECTION CELLS

PETER ZEIDMAN JOHN A. BULLINARIA

School of Computer Science, The University of Birmingham
Edgbaston, Birmingham. B15 2TT, UK

After a review of the purpose and biological background of Head Direction Cells, and existing models of them, we introduce an improved neural model that integrates visual flow information with the vestibular inputs found in earlier models. Simulation results using real video inputs are presented to explore and validate the new model, the evolutionary advantages for the proposed arrangement are noted, its utility for use in robot controllers is discussed, and suggestions are made for further work in this area.

1. Introduction

The study of *navigation* is of interest in psychology, engineering, computer science and neurobiology. To navigate an environment successfully requires the integration of multi-modal sensory information, the maintenance of an accurate world model, and the ability to localize oneself and recover from mistakes. The neurological basis of such abilities has been studied extensively, and modelling the brain's methods for performing accurate navigation may also lead to the construction of an improved generation of mobile robots.

Successful navigation has three main functional requirements:

1. A Map: An internal representation of the environment needs to be generated, to allow sensory information to be associated with localities. This is the general function of *Spatial Memory*.
2. Knowledge of Current Location: O'Keefe & Nadel (1978) described cells in rats' brains that act like a map of their environment. These *Place Cells* fire maximally when the rat is in a particular location. If moved to a different setting, the place cells spontaneously reconfigure, with each group of cells corresponding to a specific area of that environment (the "place field").
3. Heading Direction: In addition to one's map position, the heading direction is required for navigation. The neural representation of heading is via cells known as *Head Direction* (HD) Cells, and this "biological compass" is the focus of this paper. They were discovered by J.B. Ranck in 1984, and the first detailed study was published by Taube, Muller & Ranck (1990).

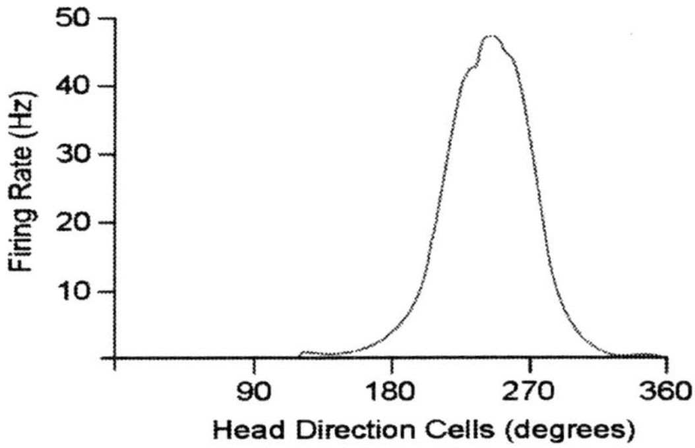

Figure 1: Typical Head Direction Cell firing profile. In this case the cell's *preferred firing direction* is 250 degrees (Sharp, Blair & Cho, 2001).

The calculation of heading direction is a neural process that translates body-centered sensory data (such as the retinal position of a visual cue) into a heading in a world-centered model. The ability of the brain to produce these location-invariant representations is an active research area and has implications both for our understanding of the brain and for the design of intelligent robots.

The following begins with an overview of the biological background behind Head Direction Cells, then introduces the sensory information used by the cells and the brain regions in which the cells are located. A review is given of the computational techniques and models that exist, and then our extended model is presented that integrates vestibular inputs with visual flow information.

2. Biological Background

2.1. *Head Direction Cells*

We have seen that navigation requires knowledge of heading, and that HD cells in the brain act as a biological compass. It is not a compass that uses magnetic fields. Instead, a combination of *ideothetic* (internally generated) and *allothetic* (externally generated) cues are used to update the heading representation.

Each Head Direction cell fires maximally for one heading direction, termed the *preferred firing direction*. Heading angles are world centered, i.e. measured from the perspective of a stationary external observer. Figure 1 shows a sample firing pattern from a single cell. Each cell's preferred firing direction is thought to be configured upon entering an environment, or given certain sensory stimuli. These directions are distributed equally around 360 degrees, and the cells operate as an integrated system: only one heading being represented at a time.

There are three major considerations for HD cell models:

- How should the HD representation be formed?
- How should the HD system track changes in its host's heading, so as to change which HD cells are firing at a given time?
- What mechanisms can reconfigure the preferred firing directions of the HD cells, so that the cells come to represent a different heading?

We proceed by examining the different sensory inputs thought to contribute to HD cell firing, consider the differences in brain areas where HD cells are formed, and then move on to the regions of the brain where HD cells are located.

2.2. *Sensory Inputs*

By combining allothetic and ideothetic cues, the HD cells estimate the host's current heading. The primary sense employed by HD cells comes from the Vestibular System. Vestibular input is generated by structures in the inner ear, and signals the horizontal acceleration of the host's head in either direction. Experiments have shown that disabling the vestibular sense removes the head direction signal (Stackman, Clark & Taube, 2002).

Vision may be divided into two categories: object (scene) recognition and optic flow. These two modalities of visual information are treated differently by the head direction system, and should therefore be considered independently.

Salient visual cues are used by the Head Direction system to set the current heading. This has been tested extensively via cue rotation experiments. Taube et al. (1990) placed a rat in a tall cylinder, with the only orientating cue being a piece of white card covering 90 degrees of the circumference. Rotating the cue by 90 degrees resulted in a near-90 degree rotation in the preferred firing direction of *every* head direction cell, emphasizing not only that the cue was tracked, but also that the cells form a distributed representation of heading.

The HD cells must also track the gradual movement of the host over time, and this is where optic flow may be used. Results from Blair & Sharp (1996) demonstrate the influence of visual flow on the HD system, and its interaction with the other senses. They constructed a cylinder to house the rat with a number of strips of white card positioned equidistantly around the cylinder's circumference as cues. The wall of the cylinder was rotated 90 degrees clockwise, with the floor stationary. This generated a conflict – visual stimuli made the rat believe it was turning counter-clockwise, while its vestibular system indicated it was stationary. Eight HD cells were recorded to discover how this sensory conflict would be resolved. Six out of eight of the cells showed no rotation in their preferred firing direction, meaning that only a quarter of cells were tricked by the moving wall. This demonstrated that, in a conflict situation, less credence is given to visual than vestibular input.

Empirical studies have shown that visual cues are used as a guide for orientation, but the lesser reliance on vision compared to vestibular sense leads to the hypothesis that optic flow is a gating mechanism for vestibular input. When in agreement, optic flow boosts the vestibular signal. When it is in contradiction, it weakens the vestibular signal but does not over-ride it.

2.3. *Brain Regions*

Head Direction cells were first discovered in an area of the brain known as the Post-Subiculum (PSc), but cells that fire as a function of the rat's heading have since been discovered in other brain areas. By understanding the purpose and interaction of these brain areas, a model of the system can be built. We now describe the relevant brain areas, with reference to suggested functions.

2.3.1. *Dorsal Tegmental Nucleus (DTN)*

Head Direction cells have not been found in the DTN, but it is considered to be a major input to the HD system, so we shall discuss it first. The majority of cells in the DTN have their activity correlated with the host's angular head velocity (Bassett & Taube, 2001). Connections are received from horizontal vestibular canals; structures that signal head rotations. These signals are thought to be used for a number of purposes, including fixating eye position, as well as input for the head direction system. Asymmetric neurons consist of 27.3% of cells recorded by Bassett & Taube (2001). These cells have a greater firing rate for either clockwise or counter-clockwise turns of the host's head. Less well understood are symmetric head-direction cells (47.7% of those recorded), which fire equally given turns in either direction.

DTN is essential to the HD system: lesion studies show that damage to DTN causes loss of the HD signal in the ADN (Stackman & Taube, 1998). It can be deduced that connections from DTN form the principal input to the HD system. Connections have been found projecting from DTN to area LMN, so LMN is probably the next area in the flow of information in the HD system.

2.3.2. *Lateral Mammillary Nucleus (LMN)*

The LMN has two major inputs: the DTN and the Post-Subiculum (PSc) which is thought to handle visual input to the system. HD cells in the LMN have preferred firing directions distributed evenly around 360 degrees. This input is vital to the HD system, and bilateral lesions cause loss of HD cell firing in area ATN, described below. HD cells in the LMN code for the host's head direction up to 40ms in the future; further in the future than HD cells in the other brain areas described here. It is therefore reasonable to assume that LMN is near the source of the HD signal. Its function is thought to be part of an Attractor-Integrator network, along with DTN. This serves two purposes:

- Attractor: Maintains the population of HD cells in such a way that they represent only one heading direction.
- Integrator: Integrates the movement of the host over time, so as to update the head direction signal accordingly.

LMN provides excitatory connections to the ATN, which is the next stage in the generation of the HD signal.

2.3.3. *Anterior Thalamus (ATN)*

The ATN contains the highest percentage of HD cells of the areas discovered. Bilateral lesions of LMN causes the ATN HD cells to lose their directional firing, implying that ATN sources its input from LMN. An interesting property of the ATN is that its firing is anticipatory – it represents the host's heading 20ms in the future. Lesions to the ATN cause area PSc to lose its directional firing. Goodridge & Taube (1997) suggested that "the AD projection may serve as a gating device that activates PoS [PSc] neurons sufficiently so that they are sensitive to the directional input from other structures". This hypothesis that ATN acts as a gating device is supported by the large percentage of HD cells it contains, and places the ATN at the centre of our model.

2.3.4. *Post-Subiculum (PSc)*

Lesion studies have shown that the PSc is not an essential component of the HD system, as it does not inhibit firing in the ATN. It is, however, required for visual cues to orientate the HD system (Goodridge & Taube, 1997). It is reasonable to assume that visual input to the HD signal is either generated by the PSc, or as suggested by Goodridge, gated by the PSc. The PSc then sends connections back to LMN, completing the flow of information. However, it is not thought that the PSc sends visual information to LMN; instead it is believed that this reaches the ATN via a separate body: the retrosplenial cortex.

3. Neural Modelling

Artificial neural network models of the HD system operate as *leaky integrators*. By building up a representation over time, whilst not placing too great a reliance on any single sensory reading, the system is tolerant to noise and error.

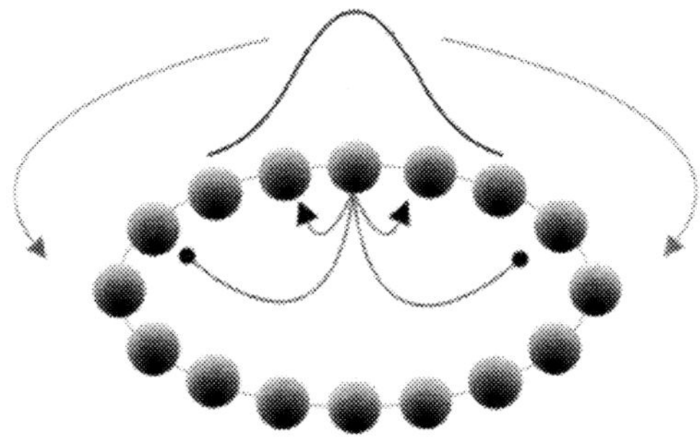

Figure 2: The Continuous Attractor Neural Network (CANN). Shaded circles represent neurons. A Gaussian "hill" of activity is shown over a number of neurons, representing their activation strength. Small pointed arrows represent local excitation; circular arrows represent distal inhibition.

3.1. *Continuous Attractors*

Attractor networks are a class of neural networks that are considered to be a good approximation to those in the HD system. Continuous Attractor Neural Networks (CANNs) consist of a set of nodes and weighted connections between them, and, depending on the implementation, there may be separate excitatory and inhibitory connections between nodes. The connection weights are pre-assigned to exhibit local cooperation and distal inhibition. This means that a node should excite its neighbours, increasing their activity level, but inhibit those further away so as to create a single peak of excitation (a "hill" of activity). The shape of this peak is determined by the weights, and its position is guided by the external inputs.

The key property of this network, that makes it useful for modelling biological systems such as Head Direction, is that the hill of activity persists after the input is removed. The network therefore acts as a short-term memory, integrating its input over time. The Gaussian hill can be shifted by supplying external input to neurons on either flank of the hill. Once input ceases, the hill comes to rest and is stable – the network is in an attractor state. The number of possible attractor states is determined by the number of neurons in the network.

Weights between nodes are a Gaussian function of their Euclidean distance:

$$w_{ij} = \exp\left(-(\theta_i - \theta_j)^2 / 2\sigma^2\right)$$

where θ_i is the preferred firing direction of neuron *i*, and σ is the width of the excitatory or inhibitory effect. The firing properties of the CANN network are governed by three time *t* dependent equations (Redish et al., 1996):

$$\tau \frac{dS_i(t)}{dt} = -S_i(t) + F_i(t)$$

$$F_i(t) = \tfrac{1}{2}(1 + \tanh(V_i(t))) \quad , \quad V_i(t) = \gamma_i + \sum_j w_{ij} S_j(t)$$

where S_i is the input received from neuron i, F_i is the probability that neuron i will fire, V_i is the firing rate of neuron i, and γ_i is a tonic inhibition term.

We next consider how such CANN networks have been used to implement HD models, and how we can use them to extend those past models.

3.2. *Existing Models*

Redish et al. (1996) modelled the relationship between ATN and PSc. Each area was represented by a CANN, consisting of an excitatory and an inhibitory layer of nodes to maintain a single hill of activation. They based this on observations that PSc represents current heading, whereas ATN represents future heading. The model replicates this by using the ATN output to guide the PSc network.

Degris et al. (2004) designed an HD representation for a mobile robot based on the interaction between areas LMN and DTN in the rat brain. It used two sensory modalities: vestibular input to the DTN attractor network, and visual signals to the LMN for correcting drift. The dynamics of their system differed from that of most other authors, in line with the observation that no recurrent connections have been discovered in the LMN. They did not use excitatory connections between LMN units, instead establishing excitatory connections to the DTN network. This in turn inhibited LMN activity according to a Gaussian profile, such that a single hill of activity was maintained.

Once an attractor network has been set up with a hill of activity, the hill needs to be translated to reflect a shift in head direction. Skaggs et al. (1995) proposed two additional sets of neurons, called *rotation cells*, to guide the hill of activity around the network. Each rotation cell receives input from both its corresponding HD cell and the vestibular system. If the input is above a set threshold, the rotation cell will fire, transmitting its activation to the HD cells. A right-rotation cell transmits excitatory signals to the HD cells on the ring to its right; conversely left-rotation cells transmit only to HD cells to their left. The effect is to excite the HD cells on the appropriate flank of the hill of activity, given the vestibular input, and shift the hill of activity.

Skaggs et al. (1995) also provided the important concept of concentric rings of neurons, that integrate heading over time using multiple sensory modalities, but it was not described formally, nor verified experimentally, so further work was needed to test it. A more detailed model was proposed by Redish et al. (1996) with dual attractor networks, one representing the ATN and one the PSc.

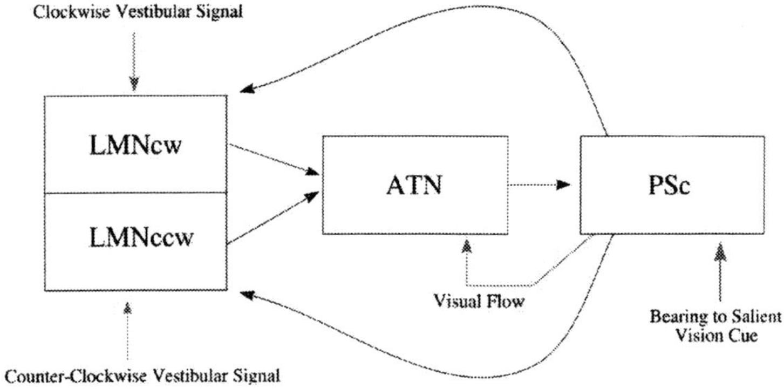

Figure 3: An overview of the extended HD model and the sensory inputs that control it. The introduction of sensory information is indicated by red/gray arrows, while the black arrows show connections between the attractor networks.

These networks were coupled by two types of connection: matching and offset. When activated by vestibular input, offset connections shifted the hill of activity, and thus the preferred firing direction of ATN cells. When vestibular input ceased, matching connections kept both networks aligned and stable.

This model generates tuning curves similar to the biologically plausible PSc and ATN cells, and the predictive relationship of ATN to PSc is faithfully replicated. Only vestibular input is considered; vision, self-motion and other senses are not mentioned. The problem with their model is that no recurrent excitatory connections have been found in the ATN, and PSc lesions do not disrupt ATN firing, so there remain doubts about the biological plausibility.

A different approach to shifting the CANN's preferred heading was used by Goodridge & Touretzky (1994). Their model was based on observations that the tuning curves of ATN HD cells are distorted as a function of the host's angular velocity. They considered the interaction between three brain areas: LMN, ATN and PSc. LMN was represented by two attractor networks, one modulated by clockwise turns, the other by counter-clockwise turns. The ATN model combined the output of the two LMN networks into a single tuning curve which was then passed to the PSc. This tuning curve adjusted the firing direction of the PSc, which was taken to be the output of the system.

3.3. *The Extended Model*

The HD model of Goodridge & Touretzky (1994) forms the basis for our model, but we need to introduce two major extensions:

- Visual flow information is transmitted from area PSc to ATN. This is in keeping with hypotheses of PSc's function in the HD system. The original Goodridge & Touretzky (1994) model only accounted for vestibular input.
- The model of PSc is further extended to allow salient visual cues to orientate the HD system. The model is therefore also made practical for use in real-world robotics applications through the combination of three sensory modalities: visual flow, object recognition and vestibular input.

Figure 3 provides an overview of the model. PSC, LMNcw and LMNccw are CANNs. Each consists a layer of excitatory neurons and a single inhibitory neuron. Weights between excitatory neurons are initialized according to a Gaussian function of the distance between them. At each epoch or time-slice, the networks are updated in turn using discrete versions of the equations given in Section 3.1, as discussed in Weiner & Taube (2005).

3.4. *Simulation Details*

Simulation software was developed to allow the model to be "placed" into a virtual cylindrical enclosure similar that used in the rat experiments of Taube et al. (1990), and to compare the results. This could later be extended to other types of behaviour testing enclosures, such as the Morris Water Maze.

Visual flow is provided to the model by a video of the enclosure. The model "sees" a 90 degree rotation clockwise, 180 degrees conter-clockwise, then 90 degrees to return to the starting position. The simulator converts this to a real-valued input parameter in the range [-1,1], with the sign indicating clockwise (positive) or counter-clockwise (negative) rotation. The visual flow algorithm is based on software by David Stavens (*Stanford Artificial Intelligence Lab*). A vector is generated for each salient feature in the image, where the bearing indicates direction of movement and length indicates the distance of that movement. Features are tracked from frame to frame and the velocity calculated.

Prior to running the model, simulated vestibular information needs to be associated with the visual stimuli. The input video is played and a numeric value manually associated with each frame. The video is then re-set and both visual and vestibular information are supplied to the model in real-time.

3.5. *Simulation Results*

3.5.1. *Tracking Vestibular Angular Velocity*

In the first experiment, simulated vestibular input complemented the visual stimulus provided to the model. The system was allowed 200 epochs to initialize

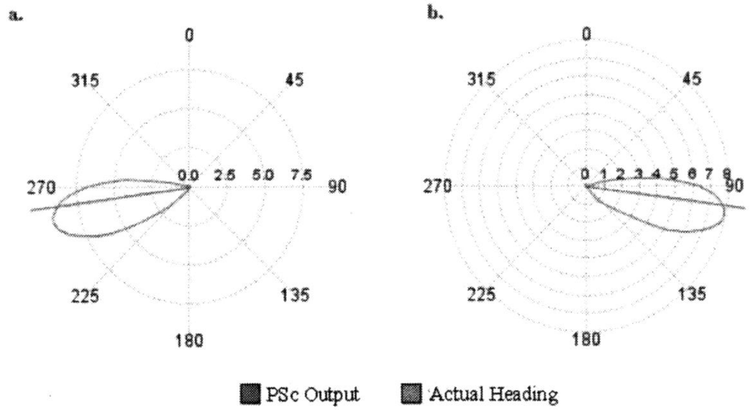

Figure 4: Directional heading graphs generated by the extended model. Chart a. shows the heading after an 85 degree cw turn, chart b. after a 180 degree ccw turn.

before the stimuli were provided. The activity of the PSc network was recorded at each epoch, and compared against the actual heading of the agent.

Figure 4 shows the model and target outputs of the CANN model at two stages of the execution: after an 85 degree clockwise rotation, and then after a subsequent 180 degree turn counter-clockwise. The network's output under-estimated the genuine rotation by about 6 degrees. This could be improved by varying the networks' connection strengths, or by using visual object recognition to periodically update the correct heading.

The velocity of a head turn in the model should reflect the angular velocity indicated by the sensory inputs. Changes in the CANN representation lagged behind the vestibular input by an average of 70 epochs.

3.5.2. *Optic Flow Integration*

A comparison was then made of the performance under sensory agreement and disparity. In the first scenario, visual flow supports the vestibular input. In the second, the visual flow signal was inverted to establish a conflict situation.

Figure 5 shows the activity of the model given matching input data for: 90 degrees clockwise, 180 degrees counter-clockwise, 90 degrees clockwise. The comparative influence of the two sensory modalities can be seen more clearly when they are made to contradict. It can be seen in Figure 6, where the direction of travel indicated by optic flow has been purposefully inverted, that vestibular sense is treated with highest priority in a conflict situation. The conflicting optic flow signal reduces the effect of vestibular input, reducing the angular velocity of the model. The amount by which this reduction occurs is dependent on the parameters of the model, and may be adjusted accordingly.

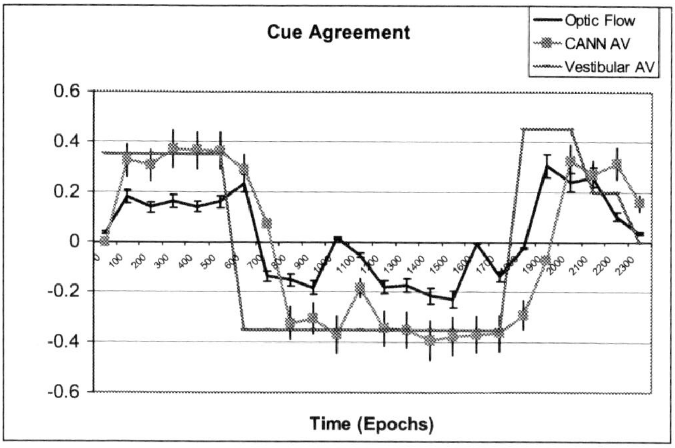

Figure 5: The angular velocity of the CANN in the presence of matching optic flow and vestibular input. (The angular velocity is scaled to lie in the range -1 to +1.)

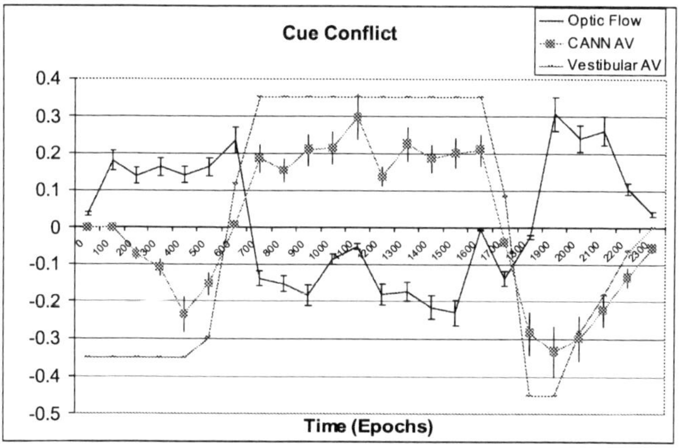

Figure 6: The angular velocity of the CANN network in a cue conflict situation.

4. Conclusions

The extended Head Direction model we have presented supports the hypothesis that visual flow acts as a gating function on vestibular inputs. The evolutionary advantage of such an arrangement is clear: visual information is far more subject to noise and misinterpretation than internally generated senses, and thus should be trusted less when disagreement occurs. The model tracks heading successfully and is a reasonable abstraction of real brain function.

The original motivation for this work was as inspiration for building robotic controllers, and the model offers two significant advantages in this application:

- The model degrades gracefully in the event of sensor damage, meaning that it will operate reasonably even if all external sensors are disabled.
- Additional senses with different modalities may be added, to increase resilience in the event of any individual sense becoming confused.

There are two aspects of the current work that could be improved upon fairly easily. First, the optic flow input used in the simulation was very noisy, largely due to a poorly illuminated input video. Further testing should be conducted under improved conditions. Second, the parameters that control the attractor dynamics could be altered so as to avoid the system underestimating its current heading. Simulated evolution by natural selection is one approach that might prove helpful here, in which populations of attractor networks with different parameters repeatedly compete and reproduce leading to improved performance.

Future work will also need to consider further the precedence of the different senses during multi-modal integration, and quantify the point at which sensory disagreement leads to confusion about orientation. In the case of implementing Head Direction models for mobile robots, the task will be to automatically set appropriate thresholds for trusting or distrusting digital sensors.

References

Bassett, J.P. & Taube, J.S. (2001). Neural correlates for angular head velocity in the rat dorsal tegmental nucleus. *Journal of Neuroscience*, **21**, 5740–51.

Blair, H.T. & Sharp, P.E. (1996). Visual and vestibular influences on head direction cells in the anterior thalamus of the rat. *Behavioural Neuroscience*, **110**, 643–660.

Degris, T., Lacheze, L., Boucheny, C. & Arleo, A. (2004). A spiking neuron model of head-direction cells for robot orientation. In *Proceedings of the Eighth International Conference on the Simulation of Adaptive Behavior, from Animals to Animats*, 255–263.

Goodridge, J.P. & Taube, J.S. (1997). Interaction between the postsubiculum and anterior thalamus in the generation of head direction cell activity. *Journal of Neuroscience*, **17**, 9315–9330.

Goodridge, J. & Touretzky, D. (1994). Modeling attractor deformation in the rodent head-direction system. *Journal of Neurophysiology*, **83**, 3402–3410.

O'Keefe, J. & Nadel, L. (1978). *The Hippocampus as a Cognitive Map*. Oxford: Oxford University Press.

Redish, A.D., Elga, A.N. & Touretzky. D.S. (1996). A coupled attractor model of the rodent head direction system. *Network: Computation in Neural Systems*, **7**, 671–685.

Skaggs, W.E., Knierim, J.J., Kudrimoti, H.S. & McNaughton, B.L. (1995). A model of the neural basis of the rat's sense of direction. *Advances in Neural Information Processing Systems*, **7**, 173-180.

Stackman, R.W. & Taube, J.S. (1998). Firing properties of head direction cells in rat anterior thalamic nucleus: Dependence on vestibular input. *Journal of Neuroscience*, **17**, 9020–9037.

Stackman, R.W., Clark, A.S. & Taube, J.S. (2002). Hippocampal spatial representations require vestibular input. *Hippocampus*, **12**, 291–303.

Taube, J.S., Muller, R.U. & Ranck, J.B. (1990). Head direction cells recorded from the postsubiculum in freely moving rats. II. Effects of environmental manipulations. *Journal of Neuroscience*, **10**, 436–447.

Wiener, S.I. & Taube, J.S. (Eds). (2005). *Head direction cells and the neural mechanisms of spatial orientation*. Cambridge, MA: MIT Press.

RECURRENT SELF-ORGANIZATION OF SENSORY SIGNALS IN THE AUDITORY DOMAIN

CHARLES DELBÉ

LEAD-CNRS UMR 5020, University of Burgundy, France

In this study, a psychoacoustical and connectionist modeling framework is proposed for the investigation of musical cognition. It is suggested that music perception involves the manipulation of 1) sensory representations that have correlations with psychoacoustical features of the stimulus, and 2) abstract representations of the statistical regularities underlying a particular musical syntax. Simulation results of two behavioral experiments investigating the interaction of sensory and cognitive components in the so-called harmonic priming task are reported.

1. Introduction

1.1. *Music perception as syntax learning and processing*

1.1.1. *Music and syntax*

Music, like language, is a highly structured domain, in which a set of principles governs the combination of discrete structural elements into sequences. These combinatorial principles can be observed at multiple levels, such as the formation of chords from musical tones, chord progressions and keys (*i.e.* the possible sets of chords). This is particularly true for the Western tonal system, where 12 pitch classes[a] are organized in sets of seven tones resulting in 12 major and 12 minor keys. In a given key, seven chords can be defined, each on a different degree of the scale. It has been shown that there exists a harmonic (*i.e.* chordal) hierarchy [1]; for example, chords built on the first (I), fourth (IV), and fifth (V) degrees of the scale (referred to as tonic, subdominant, and dominant chords, respectively) are the most important chords of the key compared to chords built on other scale degrees, the tonic chord being perceived as the most stable one. Hence, chords have particular tonal functions within a key. Remarkably, the same chord can have different musical functions

[a] referred to with the labels C, C# or D, D, D# or E, E, F, F# or G, G, G# or A,A, A# or B, and B.

depending on the context, *i.e.* the current key of a musical passage. For example, the C major chord acts as the most referential tonic in the key of C major, as a less referential dominant in the key of F major, and as an even less referential subdominant in the key of G major.

Numerous researches in the domain of music perception and cognition corroborate the hypothesis that listeners possess some knowledge about the *statistical* distribution of musical events (*e.g.,* [2]). In the case of Western tonal music, the processing of chord progressions is governed by an implicit (syntactic) knowledge of this musical idiom. Furthermore, this knowledge is acquired through mere *exposure* to a musical environment [3]; indeed, one of the most striking experimental findings concerned the very weak influence of musical expertise on the perception of harmonic structures. Naives listeners (nonmusicians) seem to qualitatively perceive musical structures in the same way that expert listeners (musicians) do (for a review of behavioral data, see [4]).

1.1.2. *The harmonic priming paradigm*

One of the strongest behavioral evidence for the syntactic processing of Western harmony comes from studies using a harmonic priming paradigm. For example, Bharucha and colleagues investigated priming effects elicited by a single chord ([5-8]). Participants heard a prime chord followed by a target chord. The prime and target chords were either harmonically closely related, or harmonically distantly related. The harmonic relatedness of the two chords was defined on the basis of their music theoretical relationship: the two harmonically related chords shared parent keys, whereas the two harmonically unrelated chords did not (chord relatedness can be represented by the so-called Circle Of Fifth, see Fig. 1, insert). For half of the trials, the pitch of one of the components of the target chord was slightly changed and the participants had to decide as quickly and as accurately as possible if the target chord was in-tune or out-of-tune. Priming effects are reflected in a bias to judge targets to be in-tune when they were related to the prime, and in shorter response times for in-tune related targets and for out-of-tune unrelated targets.

The observed processing facilitation between harmonically related chords does not only occur in case of pairs of chords. Indeed, when the local context is kept constant, global harmonic context have also been shown to influence the processing of musical events. In two studies, Bigand and Pineau [9, 10] created pairs of eight-chord sequences in which the two final chords were identical for each pair. The first six chords, however, established two different harmonic contexts, one in which the final chord was highly expected (a tonic following a dominant) and the other in which the final chord was less expected (a subdominant following a tonic). Hence, the harmonic function of the target was manipulated by varying the global context. Here again, target chords were processed more accurately and faster in the highly expected condition than in the less expected condition, indicating an effect of global harmonic context.

1.1.3. *Sensory and cognitive components in harmonic priming*

In Western tonal music, the most referential events (*e.g.*, the tonic chord) occur more often than hierarchically less important events (*e.g.*, dominant and subdominant chords): harmonic hierarchies are strongly correlated with the statistical distributions of tones and chords [2]. Hence, in a harmonic priming task, related targets have more frequency components in common with the context that do unrelated targets. It is then possible that harmonic priming may have been confounded with a form of repetition priming (repetition of frequency components). Parncutt and Bregman [11] pointed out the difficulty to compare the relative importance of top-down (syntactic) and bottom-up (sensory) processes in harmonic perception of tonal music, due to the quantitative similarity of their outcomes: the predictions made by the two approaches often correlate significantly with each other. For example, Leman ([12]) simulated the probe-tone experiments of [1] (which was taken to probe the long-term knowledge of listeners about the tonal hierarchy), using a purely bottom-up auditory model, involving only short-term memory of pitch periodicities. Actually, in the music perception domain, although the opposition between sensory processes and abstract knowledge has been investigated in few researches, the interaction between these two factors is still not clear. Nevertheless, two experimental studies were proposed to investigate the respective role of sensory and cognitive components in harmonic priming.

In Tekman and Bharucha's study [8], the target was either more psychoacoustically similar to the prime (e.g., C and E chords, having one component tone in common, see Fig. 1, left), or more closely related on the basis of harmonic convention (C and D, having one key in common). Furthermore, the Stimulus Onset Asynchrony (SOA) between the two chords was manipulated. The results revealed facilitation for psychoacoustically similar targets for a short SOA (50 ms), and facilitation for harmonically related targets after a longer SOA (> 50 ms).

Another elegant study was proposed by Bigand *et al.* [13] to disentangle sensory and cognitive processes in harmonic priming task. Here, the occurrence of the target chord in the context was manipulated. When the target acted as a subdominant, it did appear many times in the context; however, when a tonic chord was used as a target, it did not appear in the context. Hence, less expected chord shared many periodicities with the acoustical context (Fig. 1, right). The main outcome of the study suggested that the cognitive component is a fast-acting component that prevails against sensory priming at low and normal tempi (SOA=300ms), but that the effect of harmonic relatedness is reversed at fast tempi (SOA=75ms).

These two studies underline the complex interactions between sensory and cognitive components in harmonic priming tasks. The classical model of this task, MUSACT ([3, 13]), predicts the inversion effect in case of chord pairs, but fails to predict behavioral results of Bigand *et al.* The present simulation study argues that these complex phenomenons could be explained in the light of the interactions of 1) adequate pitch-related representations of the stimuli and their corresponding echoic memory traces and 2) abstract syntactic knowledge.

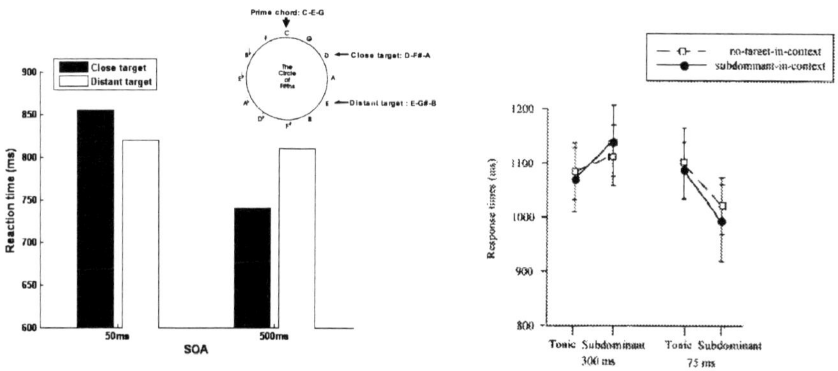

Figure 1. Results of two harmonic priming tasks. Left : reaction times on the target for chord pairs [8]. All chords were major chords, being the principal chord of a key from the circle of fifths (insert). In this case, the prime is a C-major chord, the close target is a D major chord, and the distant target is an E major chord. Right : reaction times for tonic (related) versus subdominant (less-related) targets in long context [12], at two different tempi. SOA=Stimulus Onset Asynchrony.

2. Simulations

2.1. *The present model*

The model of harmonic perception proposed in the present paper involves the formation and processing of different representations of sound patterns (Fig. 2). Firstly, these inputs are transformed in neural rate-coded representations by an auditory peripheral model. This auditory nerve activity is then analyzed by means of a temporal pitch model, giving periodicity pitches estimation, labeled $\vec{p}^{(t)}$. Second, a short-term memory model stores a leaky-integrated trace, labeled

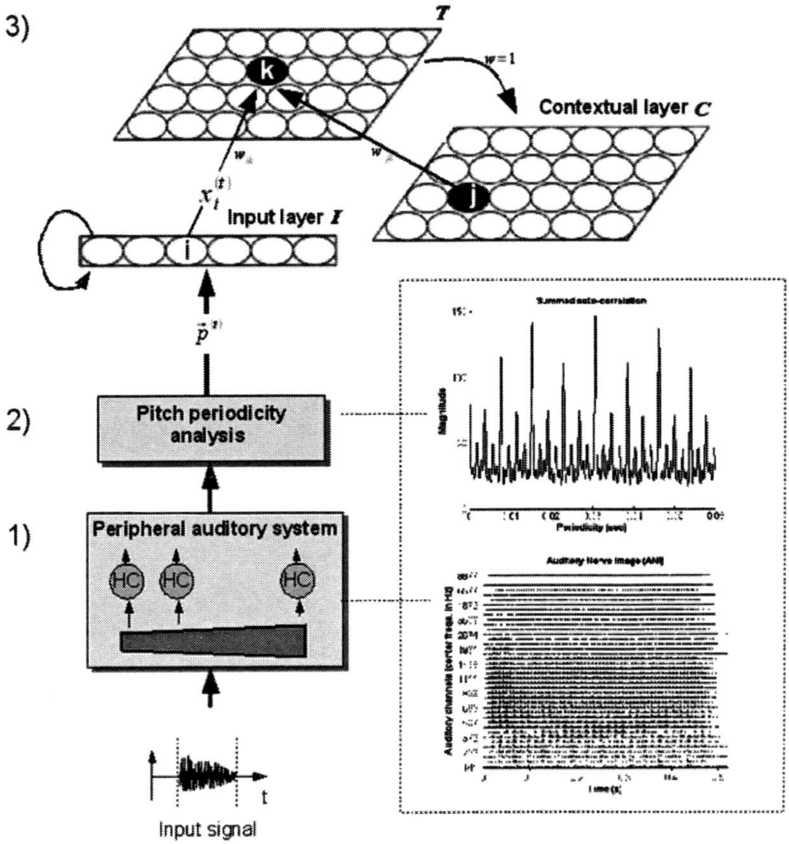

Figure 2. Main components of the proposed model. Acoustic input signal is first processed by a model of the peripheral auditory system, giving an estimation of the probability of firing of a number of hair-cell neurons (1). Common periodicities among the simulated fibers are calculated (2), and fed into the input layer of a recursive map (3). HC = Hair-Cell.

$\vec{x}^{(t)}$, of the output of the pitch model. Finally, the state of this echoic memory is fed into a connectionnist model of sequence learning. We will now present in detail the different modules of the model.

2.1.1. *The auditory peripheral model*

There are two basic ideas about how pitch information of a musical signal can be encoded in patterns of neural discharge: coding by spatial patterns of neural excitation vs. coding by temporal patterns in spike trains. Spatially distributed spectral representations are supported by the finding that particular populations of auditory neurons are tuned to a particular frequency range [14]. The profile of average discharge rate across these tonotopic auditory frequency maps forms a neural representation of the stimulus power spectrum. However, temporal coding of pitch is supported by many physiological studies, in animal [15], as well as in human [16, 17]. The temporal representation is based on the distributions of time intervals between the spikes generated by the auditory neurons, in response to the basilar membrane displacement. It has been established that strong correspondences exist between inter-spike interval statistics over the auditory nerve fibers and the pitches produced by a wide range of complex tones [17]. Furthermore, other aspects of pitch can be directly explained in term of population-interval distributions at the level of auditory nerve: the pitch of the missing fundamental[b], pitch equivalence of stimuli with different power spectra (*i.e.* different timbres), and pitch produced by unresolved harmonics.

The pitch representation adopted in this modeling study involves both spatial and temporal coding. To this end, the peripheral auditory model of Van Immerseel and Martens [18], adapted by Leman [12], was used. The first stage of the model transforms an acoustical signal into a set of patterns of neural firing rate-codes in 40 auditory nerves. This first step simulates the filtering processes of the outer and middle ear, the resonance of the basilar membrane and the temporal dynamics of the auditory neurons (*i.e.* the hair cells). The resulting patterns of firing probabilities are then processed by a temporal model of the perception of the pitch of complex sounds. Specifically, a periodicity analysis, by mean of an autocorrelation function[c], is performed for each 40 output channels of the auditory model, using a window of 60 ms. The 40 resulting periodicity patterns are summed up at each time step (of 60 ms), giving a single autocorrelation pattern for each window (note that for the simulations reported here, a single mean pattern $\vec{p}^{(t)}$ was calculated for the duration of each chord). This pattern has very interesting properties; indeed, it represents common periodicities among the auditory neurons in a given frequency range. Given an incomplete harmonic complex tone, its analysis in terms of a common periodicity of the auditory resolved harmonics will point to the fundamental of the harmonic complex. Indeed, the summed autocorrelation function shows peaks at the frequency components of the sound. Moreover, peaks at subharmonics, integer multiple of the component's period, are present. If the

---

[b] An harmonic complex sound in which the fundamental frequency (F0) is physically missing still evokes a pitch at that frequency.

[c] This function compares a signal with different temporally delayed versions of it, in order to find the delay which produces the higher correlation value between these two versions. It indicates periodicities, or repetitions, present in the signal.

stimulus is a harmonic complex, then all components have a common subharmonic at the fundamental. When all the periods corresponding to all of the subharmonics are summed together in the 40 autocorrelation patterns, the most common periods produces a peak at the fundamental frequency and its multiple (*i.e.* its subharmonics). Such pitches are called « periodicity pitches » because the fundamental pitch reflects the periodicity of the recurring time pattern that is associated with the whole harmonic complex. Studies by Langner and collaborators [19] support the existence of representations of periodicity pitch in the auditory system.

2.1.2. *The echoic memory*

Sensory memory is usually defined as low-level memory that occurs without consciousness, abstraction, or syntactic processing. In other words, a sensory memory processes information in a purely bottom-up manner. In this study, echoic memory is simulated at the input layer I . The echoic memory layer takes a pitch vector $\vec{p}^{(t)}$ as input and returns its leaky integrated trace $\vec{x}^{(t)}$ or echoic trace as output. This trace is integrated so that at each time step, the new trace is calculated by taking a certain amount of the old trace, which is then added with the new incoming vector. The specified echo θ (in sec.) defines the amount of context that is taken into account, in function of the sampling rate *sr* (in Hz) :

$$\vec{x}^{(t)} = \vec{p}^{(t)} + \vec{p}^{(t-1)} \cdot 2^{-(\frac{1}{sr \cdot \theta})} \tag{1}$$

2.1.3. *Sequential learning and the Recursive Self-Organizing Map*

Musical pitch is represented in the auditory cortex. This cerebral structure provides an excellent example of a somatotopic processing array, as it shows tonotopic (pitch-dependent) organization in multiple processing stages. In an effort to model the somatotopic maps found in the cerebral cortex, Kohonen [20] developed the Self-Organizing Map (SOM). Like in an animal's cerebral cortex, the SOM will form regions on its two-dimensional surface in which similar features are mapped relatively close to each other while features that are more distant are mapped relatively far away from each other. A conventional self-organizing map is formed of an input layer and a topological layer, which is composed of N neurons positioned on a rectangular, usually a 1 or 2 dimensional grid. The neurons are connected to adjacent neurons by a neighborhood relation defining the topological structure of the map. In the 2-dimensional case, the neurons of the map can be arranged either on a rectangular or a hexagonal lattice. Furthermore, if the sides of the map are connected to each other, the global shape of the map becomes a cylinder or a toroid. In the original algorithm, each neuron $k \in 1 \ldots N$ compares its prototype-vector w_k to an input-vector $\vec{x}^{(t)}$, where t is a time index. This comparison is generally based on the Euclidean distance. For a given input vector $\vec{x}^{(t)}$, the error of the neuron k is defined by :

$$E_k = \left\| \sum_i x_i^{(t)} - w_{ik} \right\|^2 \tag{2}$$

The winner neuron W, *i.e* the neuron of the topological layer that best represents the input vector, is the one that minimizes this error:

$$E_W = \min_{k \in 1 \ldots N} E_k \tag{3}$$

The learning algorithm then adapts the connection weights of the winner neuron and its neighbors, in direction of the input vector:

$$\Delta w_{ik} = \eta \cdot h_{kW}^{(t)} \cdot (x_i^{(t)} - w_{ik}) \tag{4}$$

where W is the winner neuron index, η is a learning rate and h_{kW} a so-called neighborhood function, which decrease with the distance between neurons k and W on the topological layer. This neighborhood function, which defines the connection strength between adjacent neurons, is classically Gaussian:

$$h_{kW}^{(t)} = \exp\left(\frac{d(k,W)^2}{\sigma(t)^2} \right) \tag{5}$$

where $d(k,W)$ is the distance between neurons k et W on the map, and σ is the spread term of the Gaussian function. During learning, the spread of the Gaussian neighborhood function usually starts very wide, which allows the network assigning different regions of input space to different regions on its own two-dimensional surface. This neighborhood decreases until it includes only one neuron, at which time each neuron begins to specialize on certain regions of the input space. At the end of training, neighbors neurons specialize in neighbors regions of the input space, so the resulting network can be thought of as a somatotopic map relating the output space to the input space.

Although it has produced very good results with static inputs, it is often pointed out in the literature ([21, 22]) that the standard SOM is not designed for time-domain processing. As underlined earlier, music perception involves sequential processes. Extensions of the SOM that learn temporal dynamics have been proposed. These models show the ability to maintain state and memory based on past input while engaging in self-organizing learning and context recognition, making them strong candidates for modeling processes involved in music cognition.

A temporal adaptation of the Self-Organized Map of Kohonen, proposed by Voegtlin [21], was used to learn periodicity pitches in sequence. This recurrent self-organizing map learns to recognize *context,* defined as the series input data from the beginning up to the present, in a temporal sequence. Here, the activity of a topological neuron k depends on the current input vector and on the vector of activities $y^{(t-1)}$ of the topological layer at the previous time:

$$y_k^{(t)} = \exp\left(-\alpha \sum_i \left\| x_i^{(t)} - w_{ik} \right\|^2 - \beta \sum_j \left\| y_j^{(t-1)} - w_{jk} \right\|^2 \right) \tag{6}$$

Note that this differs from the traditional SOM in that neural activation depends on the network's entire previous output as well as the current state of the input layer (here, the echoic layer). The weights α and β can be used to determine the relative importance of external input versus self-feedback. The recurrent connections are central to the network's implicit memory and ability to represent temporal sequences. In the Recursive SOM, the original learning rule of Eq. 4 is applied to both connections sets as follows:

$$\begin{aligned} \Delta w_{ik} &= \eta \cdot h_{ik}^{(t)} \cdot (x_i^{(t)} - w_{ik}) \\ \Delta w_{jk} &= \eta \cdot h_{jk}^{(t)} \cdot (y_j^{(t-1)} - w_{jk}) \end{aligned} \tag{7}$$

2.2. *The training corpus*

Rather than using sequences generated from a statistical estimation of the transitional probabilities between chords of a given musical corpus, the ecological modeling approach of this study permitted to use real tonal music. 10 Bach's chorals were used to train the model[d]. The chorals were approximately one minute length, and played with Shepard tones[e], digitized at the standard CD quality (44.1 kHz, 16 bits resolution), which corresponds to the most frequently used quality in music perception research. Each choral was transposed in all possible keys, giving a training corpus of 120 sequences, for a total of 6840 chords. In order to reduce the amount of training data, vectors of pitch

---

[d] MIDI files were obtained from http://www.jsbchorales.net/.

[e] Chords components were generated by octave-spaced intervals within a bell-like spectral envelope.

periodicities ($\vec{p}^{(t)}$) were calculated for each musical beat, and then concatenated to obtain a sequence of isochronous pitch vectors for each choral.

2.3. *The training parameters*

It was noticed earlier that the proposed model uses discrete time quantization, rather than continuous variation of the output of the auditory model. The sampling rate *sr* of the sensory memory of Eq. 1 was set to 2 Hz, and the echo θ to 0.25 sec., what would correspond to a chord length of 0.5 sec. in the continuous domain.

15 Recursive SOM were trained. Each input layer was composed of 109 neurons, corresponding to the periodicity pitch vectors dimension. The topological layers were constituted of 14x16 neurons. Map weights were randomly initialized between 0 and 1. The entire training corpus was presented 20 times to the networks, in a random order, the context being reset to 0 between each sequence. During learning, the neighborhood spread σ and the learning rate η from Eq. 5 were linearly decreased from 7 to .1 and .2 to .01, respectively. The free parameters α and β from Eq. 6 were fixed and set to 0.5.

2.4. *Results*

2.4.1. *Topographic representation of periodicity informations*

In order to test the topological ability of the model, activities in response to 12 prototypical major chords were recorded, without considering the position of the chords in the sequences. Fig. 3 shows that adjacent chords on the Circle of

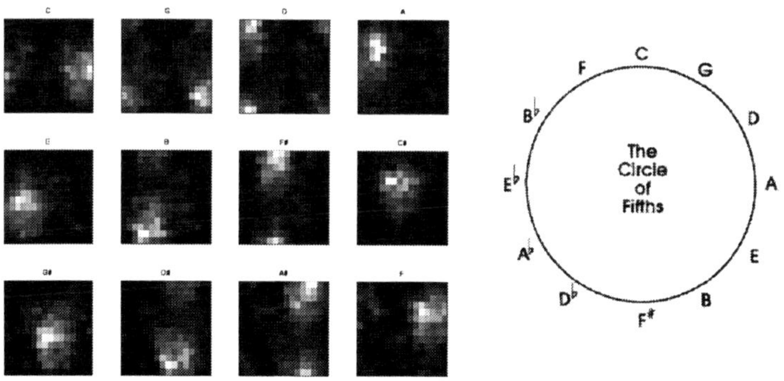

Figure 3. Left : example mean map activities in response to 12 prototypical major chords appearing in harmonic sequences. Clear areas indicate high activity. Note that right and bottom sides are connected to left and up sides, respectively. Right : the circle of fifths. adjacent chords on this circle are mapped onto adjacent neurons on the topological layer.

Fifths are mapped onto adjacent neurons on the topological layer; the self-organizing learning algorithm has thus achieved a realistic tonotopic mapping of the periodicity informations.

2.4.2. *Effect of sensory component*

The main goal of this paper was to simulate behavioral data from [8] and [13], where effects of cognitive versus sensory components in priming tasks were investigated. Fig. 4 illustrates mean activity of the winner neuron in response to the target chords, with pairs of chords used by [8] (left), and stimuli from [13] (right).

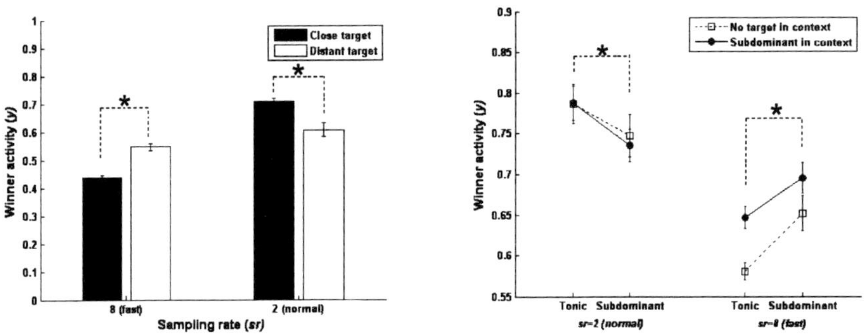

Figure 4. Simulation results for [8] and [13] (compare to Fig. 1). Note that in the case of [13], Shepard tones were used instead of piano tones, all other things being identical.

The simulation results closely resemble those of the two experimental studies of interest; in the condition where the sampling rate *sr* was set to 2 (*i.e.* at a normal tempo), activity of the winner neuron was higher for close targets than for distant ones. Conversely, when the sampling rate *sr* was set to 8 (*i.e.* at a fast tempo), the opposite pattern was found, suggesting that in this specific case, sensory component prevails over syntactic knowledge in the resulting map activity.

3. Conclusions

These modeling results support the recent hypothesis of a mutual interaction of the sensory and cognitive components in harmonic priming studies.

It has been necessary to model other brain regions using simpler, non-neural models that provide realistic inputs to the self-organizing map. Despite the simplifications, all the components of the proposed model are biologically plausible. These simulated inputs from the auditory pathway allowed the Recursive SOM to extract useful representations of both statistical properties of the harmonic sequences, and of psychoacoustical similarities between chords of these sequences.

Furthermore, these two factors not only influenced the simulated perceptual task, but they also shaped the internal representations developed by the recurrent connectionnist model. Indeed, preliminary results not reported here suggest that the strength with which sensory signals and contextual signals interact during learning determines the type of topology realized in the topographic map (*i.e.*, a spatially or temporally defined signal topology). Future work should explore the developmental trajectory of the map representations, as behavioral data from infants exist in the literature.

Acknowledgments

The author is grateful to Robert French and Emmanuel Bigand for helpful support.

References

1. C.L. Krumhansl and E.J. Kessler, Tracing the dynamic changes in perceived tonal organization in a spatial representation of musical keys. *Psychological Review*, **89**(4), 334-368 (1982).
2. C.L. Krumhansl, *Cognitive foundations of musical pitch*, New York: Oxford University Press (1990).
3. B. Tillmann, J.J. Bharucha, and E. Bigand, Implicit learning of tonality : A self-organizing approach. *Psychological Review*, **107**, 885-913 (2000).
4. E. Bigand and B. Poulin-Charronat, Are we "experienced listeners"? A review of the musical capacities that do not depend on formal musical training. *Cognition*, **100**(1), 100-130 (2005).
5. J.J. Bharucha and K. Stoeckig, Reaction Time and Musical Expectancy: Priming of Chords. *Journal of Experimental Psychology: Human Perception and Performance*, **12**(4), 403-410 (1986).
6. J.J. Bharucha and K. Stoeckig, Priming of chords: Spreading activation or overlapping frequency spectra? . *Perception & Psychophysics*, **41**(6), 519-524 (1987).
7. H.G. Tekman and J.J. Bharucha, Time course of chord priming. *Perception & Psychophysics*, **51**(1), 33-39 (1992).
8. H.G. Tekman and J.J. Bharucha, Implicit Knowledge Versus Psychoacoustic Similarity in Priming of Chords. *Journal of Experimental Psychology: Human Perception and Performance*, **24**(1), 252-260 (1998).
9. M. Pineau and E. Bigand, Influence du context global sur l'amorçage harmonique. *L'Année Psychologique*, **97**, 385-408 (1997).
10. E. Bigand and M. Pineau, Context Effects on Musical Expectancy. *Perception & Psychophysics*, **59**, 1098-1107 (1997).

11. R. Parncutt and A.S. Bregman, Tone Profiles Following Short Chord Progressions: Top-Down or Bottom-Up ? . *Music Perception*, **18**(1), 25-57 (2000).
12. M. Leman, An auditory model of the role of short-term memory in probe-tone ratings. *Music Perception*, **17**(4), 435-463 (2000).
13. E. Bigand, B. Poulin-Charronat, B. Tillmann, F. Madurell, and D.A. D'Adamo, Sensory versus cognitive components in harmonic priming. *Journal of Experimental Psychology: Human Perception and Performance*, **29**, 159-171 (2003).
14. T.M. Talavage, M.I. Sereno, J.R. Melcher, P.J. Ledden, B.R. Rosen, and A.M. Dale, Tonotopic organization in human auditory cortex revealed by progressions of frequency sensitivity. *Journal of Neurophysiology*, **91**, 1282-1296 (2004).
15. D. Bendor and X. Wang, The neuronal representation of pitch in primate auditory cortex. *Nature*, **436**(7054), 1161-1165 (2005).
16. P.A. Cariani and B. Delgutte, Neural correlates of the pitch of complex tones. I. Pitch and pitch salience. *Journal of Neurophysiology*, **73**(3) (1996).
17. P.A. Cariani and B. Delgutte, Neural correlates of the pitch of complex tones. II. Pitch shift, pitch ambiguity, phase-invariance, pitch circularity, and the dominance region for pitch. *Journal of Neurophysiology*, **76**(3) (1996).
18. L. Van Immerseel and M. J.P., Pitch and voiced/unvoiced determination with an auditory model. *Journal of the Acoustical Society of America*, **91**, 3511-3526 (1992).
19. G. Langner, C.E. Schreiner, and U.W. Biebel, *Functional implications of frequency and periodicity coding in auditory midbrain*, in *Psychophysical and Physiological Advances in Hearing*, A.R. Palmer, et al., Editors, Whurr: London 277-285 (1998).
20. T. Kohonen, *Self-Organizing Maps*, Berlin: Springer-Verlag (1984).
21. T. Voegtlin, Recursive Self-Organizing Maps. *Neural Networks*, **15**(8-9), 979-991 (2002).
22. M. Strickert and B. Hammer. *Self-Organizing Context Learning*. in *European Symposium on Artificial Neural Networks (ESANN 2004)*,. 2004.

RECONSTRUCTION OF SPATIAL AND CHROMATIC INFORMATION FROM THE CONE MOSAIC

DAVID ALLEYSSON

Laboratoire de Psychologie et NeuroCognition LPNC, CNRS UM5105
Université Pierre-Mendès France,
Grenoble, France

BRICE CHAIX DE LAVARENE

Grenoble Image Parole Signal Automatique, GIPSA-Lab, CNRS UMR5216
Université Joseph Fourier
Grenoble, France

MARTIAL MERMILLOD

Laboratoire de Psychologie Sociale et Cognitive, CNRS UMR6024
Université Blaise Pascal
Clermont-Ferrand, France

In this paper we present a neural network method for reconstructing a colour image from a mosaic a chromatic samples, arranged either in a regular or random manner. This method could be interpreted as a biological plausible method for the human visual system to reconstruct spatial and chromatic information from the random mosaic of cones in the retina.

1. Introduction

The human visual system has three kinds of cones for acquiring the colour in the visual environment in photopic (daylight vision) conditions. The cones have three spectral ranges of sensitivity and they are called L, M and S for their preferential sensitivity to Long, Middle and Short wavelength. Contrary to a colour image where three colours are sampled at each spatial position, in the human visual system, cones form a single mosaic where they are arranged randomly [1]. Consequently in the human retina only a single chromatic value is sampled for each spatial position at a time. This chromatic value corresponds to the type of cone L, M or S, which is present at the corresponding position in the retina. One can ask how the visual system is able to give us the sensation of colour and colour shading from a mosaic of chromatic samples. Moreover we can ask whether the random nature of the arrangement of these chromatic samples helps or diminishes our ability to perceive colour.

Another example of a mosaic of chromatic sampling is the digital camera. Actually, in most digital cameras today the acquisition of the colour image is done through a single sensor, which is covered by an array of chromatic filters. The so-called Colour Filter Array (CFA) gives the possibility for the sensor to discriminate between colours because the filter sensitivities cover three different ranges, usually Red, Green and Blue (RGB). Thus, in digital cameras we have to reconstruct three chromatic samples at each spatial position from an image with a single chromatic sample. This operation is called demosaicing [2]. In digital camera, the most used pattern is the Bayer CFA from the name of its inventor and consists of a regular pattern (Figure 4 (a)).

There is no real evidence that the human visual system reconstructs explicitly three colours information per spatial position such as the demosaicing process in cameras does [3]. Nevertheless, the spatial and chromatic information should be known at every position in the visual field to allow human to perceive shades of colours in natural scenes. Thus, we may suppose that even if the reconstruction of a colour image with three components does not occur in the visual system, this kind of representation of colour in an image can correspond to the information the human visual system needs to allow us to perceive colour. Thus, by extension, we suppose that the goal of the human visual system for colour perception could be compared to the ability to reconstruct three chromatic samples from a mosaic.

This article presents an approach where a neural network is used to reconstruct the missing colour information from a mosaic image. The method is applied on a regular and on a random arrangement of chromatic samples. The use of a neural network gives to the algorithm a biologic plausibility because the reconstruction is then given by a combination in a local neighbourhood of the existing pixels. This operation is plausible as a dendritic communication between neurons in the visual system.

2. Luminance-chrominance decomposition

The fact that the human visual system is sensitive to both achromatic spatial information and chromatic spatial information independently is well known since long ago [4]. This could be due to the way we perceived objects with their own colour automatically segregated from the shading of reflective light coming onto them.

There is no general consensus on the way the achromatic component is calculated from the chromatic component sampled by the retina. Actually, several decompositions were defined following the criterion used. The CIE recommends the use of the normalised function V(λ), which was measured on human subjects, to be the luminance visibility function.

The representation of a colour image into its achromatic and chromatic components is also interesting from a point of view of digital image processing. The following Figure 1 illustrates a decomposition of a colour image with its R, G and B values into luminance and chrominance components.

(a) (b) (c) (d) (e) (f)

Figure 1: Illustration of the luminance - chrominance decomposition (a) RGB image (b) Red component (c) Green component (d) Blue component (e) luminance (f) chrominance.

We illustrate on Figure 1 the fact that the interpretation of the colour shade and intensity level of an image is easier in luminance and chrominance decomposition than in the RGB decomposition. As an example, consider the plane front-end. From Figure 1(e) and 1(f) it is clearly a yellow uniform colour with a shade of intensity from the top to the bottom. However, the same interpretation from Figure 1(b)-1(d) is difficult.

3. Model of spatio-chromatic sampling by a mosaic of chromatic samples

Let us define a colour image I with its three chromatic components, Red, Green and Blue, as follows:

$$I = \{C_i\}, i \in [R, G, B] \qquad (1)$$

We can formalize the sampling of a colour image through a mosaic by the following equation:

$$I_{mosaic} = \sum_i m_i(x,y)C_i(x,y) \tag{2}$$

Where m_i are the modulation functions, which take value 1 if the colour i is present at position (x,y) and 0 otherwise. Each modulation function m_i defines a submosaic of the chromatic samples of colour i. The modulation functions are specific to the arrangement of chromatic samples on the mosaic. But these functions can always be rewritten as a constant part plus a fluctuation part. Let the constant part called p_i be the mean value of the submosaic m_i. We may write:

$$m_i(x,y) = p_i + \tilde{m}_i(x,y) \tag{3}$$

In that case, the constant part p_i corresponds to the proportion of each colour sample type in the mosaic. The fluctuation part $\tilde{m}_i$ takes the positive value $(1\text{-}p_i)$ in presence of the colour sample and negative values $-p_i$ elsewhere. With equation (3) we can rewrite equation (2) as follow:

$$I_{mosaic} = \underbrace{\sum_i p_i C_i(x,y)}_{Lum} + \underbrace{\sum_i \tilde{m}_i(x,y)C_i(x,y)}_{Chr} \tag{4}$$

Equation (4) shows that the mosaic image is in fact a sum of the luminance and the chrominance of the original image. The chrominance is actually subsampled and multiplexed in the mosaic image. The detail of the model can be found in Chaix et al. [5].

Figure 2 illustrates the composition of luminance and chrominance in mosaic images. In this simulation we used a mosaic image Figure 2(a) represented as a grey level image. We subtracted the luminance image defined in Figure 1(e) to the image in Figure 2(a) resulting in Figure 2(b). Then we demultiplexed the image on Figure 2 (b) by regrouping samples of the same colour sensitivity on their respective colour plane, either R, G or B. This allows representing that image in colour. Then we interpolate image on Figure 1 (c) resulting in image on Figure 2 (d). By comparing image on Figure 2 (d) and image on Figure 1(f) we can conclude that a mosaic image is a sum of the luminance image plus the subsampled and multiplexed chrominance image.

(a) (b)

(c) (d)

Figure 2: Representation of luminance and chrominance in a mosaic. (a) The mosaic image displayed in grey levels. (b) Image in (a) where luminance in Figure 1(e) was removed. (c) demultiplexing of image (b) by isolating chromatic samples according to the mosaic arrangement. (d) Interpolation of image (c).

This decomposition does not depend on the arrangement of colour samples in the mosaic. In the case of a non-periodic arrangement, the demultiplexing is done the same way, following the position of each colour classes in the mosaic. Nevertheless, the interpolation of the chrominance could be more complex because the neighbourhood of colour changes from place to place in the mosaic.

The model of decomposition in luminance and chrominance allows interpreting the Fourier spectrum of the mosaic image [6]. In Figure 3 (b) we can identify nine different regions where the energy of the Fourier transform is localized. The region in the centre corresponds to the luminance signal whereas the regions in the border correspond to chrominance. In the case of a random arrangement of chromatic samples, the chrominance is no longer localized, but is rather spread over the whole Fourier domain as shown in Figure 3 (d). The localization of luminance and chrominance in the Fourier spectrum has been used to reconstruct colour image from the mosaic [6]. But this method does not

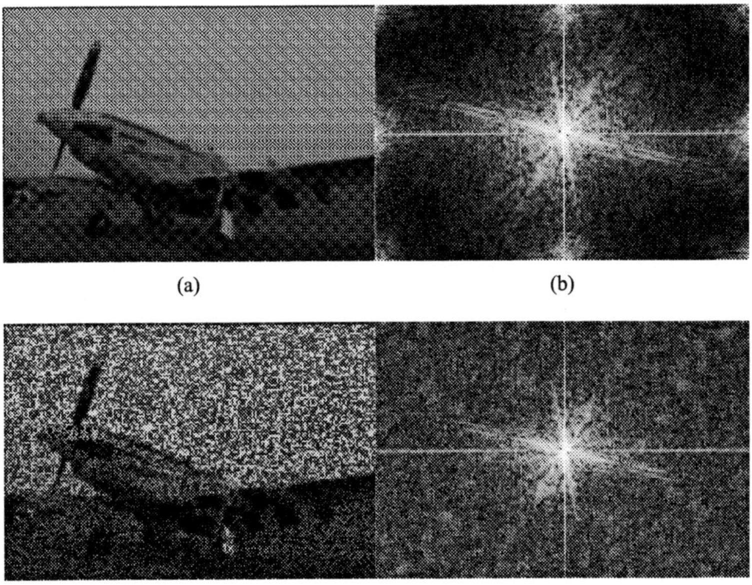

(a) (b) (c) (d)

Figure 3: Fourier representation of the mosaic. (a) The Bayer mosaic (b) The Fourier transform of the Bayer mosaic. (c) A mosaic with random arrangement of colour samples (d) The Fourier transform of the random mosaic.

apply in the case of random arrangement of chromatic samples in the mosaic. It should be found another method for the interpolation of chromatic samples in the case of random arrangement. Here we propose the use of a neural network.

4. Method

For simulating the process of sampling by a mosaic we used an image database[1] which comprises 24 colour images provided by Kodak. The images are originally of resolution 768x512 pixels. To reduce the amount of data we apply a subsampling of each individual image by a factor 4 in horizontal and vertical direction reducing the resolution to 192x128. To avoid aliasing, we apply the following low pass anti-aliasing filter before doing the subsampling.

$$f = 2^{-16} g' g \text{ with } g' = \begin{bmatrix} 1 & 4 & 10 & 20 & 31 & 40 & 44 & 40 & 31 & 20 & 10 & 4 & 1 \end{bmatrix} \qquad (5)$$

The colour image database has three chromatic samples per spatial location. We remove chromatic samples accordingly to the arrangement in the mosaic to

[1] http://www.cipr.rpi.edu/resource/stills/kodak.html

compose the input image. We then construct training and testing vector, using 12 images each.

To construct the vectors we use a neighbourhood of size 6x6 in the colour image as output vector of the neural network and the corresponding neighbourhood of size 12x12 in the mosaic image as input vector. Thus the neural network has the goal of reconstructing the three colours of 6x6 pixels from a neighbourhood of 12x12 pixels in the mosaic. We use a two layers neural network with an input layer and an output layer fully connected an initialized with random weights of mean zeros and variance 0.1. We use Lens (Light Efficient Network Simulator) [7] to perform the simulation. Once the network has converged, we use the output of the network given for the test set to reconstruct colour image. We can then compare the original colour image with the image reconstruct from the output of the neural network to test the quality of the reconstruction.

In a first experiment, we use either a Bayer or a random arrangement of size 6x6 as shown in the following Figure 4 (a) and (b) to construct the mosaic. It should be noted that we can not use a complete random arrangement for the chromatic samples because in this case the size of the input and output would be the same as the number of pixels in the image. To be able to train the network it is important that the position of chromatic samples was the same from a vector to another. Thus, we use a random pattern of size 6x6 that we tile along the image.

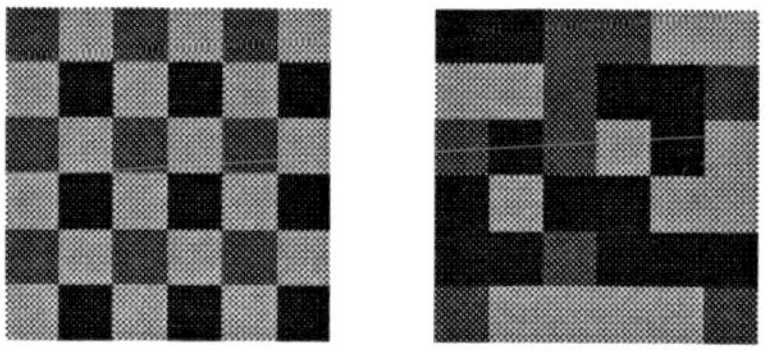

(a) (b)

Figure 4: (a) Representation of the Bayer mosaic on size 6x6 (b) Representation of the random mosaic of size 6x6 used for simulation.

In a second experiment, we use the neural network to estimate the luminance from the mosaic. Thus, the output vector is composed of a neighbourhood of 6x6 pixel of the luminance computed from the colour image.

5. Results

The result for the first experiment is given in Table 1 and Figure 5. The network converges quite quickly to a configuration that allows the reconstruction of

Table 1. PSNR between original image and image reconstruct from the mosaic by the neural network using either a Bayer or random mosaic.

	Bayer mosaic				Random mosaic			
Image	Red	Green	Blue	Total	Red	Green	Blue	Total
13	28.0667	26.9426	28.1920	**29.4156**	27.9650	26.9115	28.0575	29.2346
14	29.0981	30.4788	29.7122	**27.6237**	29.0862	30.5663	29.6630	27.5843
15	28.4966	29.5165	29.0924	**27.2336**	28.3348	29.3490	28.7703	27.1830
16	28.5257	29.7463	28.9645	**27.2510**	28.2301	29.5722	28.4295	27.0551
17	29.0636	30.4281	29.3598	**27.8071**	28.9417	30.2747	29.0963	27.8046
18	29.7367	29.4420	29.9980	**29.7883**	29.4121	29.2422	29.5339	29.4657
19	29.5661	30.4895	30.4456	**28.1836**	29.4003	30.3338	30.2847	28.0090
20	27.8741	27.8664	28.4647	**27.3612**	27.5003	27.6041	27.8171	27.1102
21	29.5636	30.0708	29.7752	28.9273	29.5528	29.9713	29.6641	**29.0716**
22	28.7614	29.4025	28.6200	28.3312	28.7111	29.6051	28.1625	**28.4934**
23	27.9490	28.1049	28.4094	**27.3959**	27.7253	27.9033	28.1576	27.1757
24	29.2061	30.3766	28.8397	**28.6059**	29.1300	30.3878	28.6826	28.5498

colour. The following table (Table 1) shows the PSNR (Peak Signal to Noise Ratio) between the original colour image and the reconstructed image from the output of the neural network.

The following Figure 5 shows an example of the reconstruction of colour image from the mosaic.

(a) (b) (c)

Figure 5: Examples of reconstruction of a colour image (a) using the Bayer mosaic (b) or random mosaic (c).

We show that the neural network is able to learn how to reconstruct the colours from the mosaic of chromatic samples in a case of regular or random arrangement of chromatic samples. The result is not perfect and it remains some error in the reconstruction such as a blurring and some false colour. However we have not tested it extensively yet and it is possible that the structure of the network or the algorithm for convergence could be improved. We also show that the global level of luminosity is not exactly reconstructed. This is certainly due to the bias used in the neural network. Finally, although the Table 1 shows that the PSNR is often higher in the Bayer case than in the random case, the images show that the random case could be better when a high frequency pattern is under consideration. This can be shown for example on the window image of Figure 5, where the false colours are reduced in the random case. This could be due to the property of the random arrangement of chromatic samples to reduce the coherence of the error of reconstruction and hence to decrease its visibility. The result for the second experiment is shown in Table 2.

Table 2. Result for the second experiment.

Image	Bayer	Random
13	**29.3712**	29.2042
14	30.9713	**31.1051**
15	**29.5052**	29.3472
16	**29.6361**	29.2351
17	**29.9640**	29.7346
18	**30.6322**	30.3208
19	**31.1623**	30.9672
20	**29.2731**	28.8540
21	**30.4569**	30.4035
22	**29.7240**	29.5396
23	**29.1436**	29.0356
24	29.8412	**29.9614**

Table 2 shows that the luminance is better reconstructed than R, G and B channel in term of objective PSNR criterion. This can be a new argument why the human visual system preferentially uses a luminance and chrominance approach than direct colour processing.

We have also tried to train a neural network with the chrominance information to see if the network was able to interpolate the chrominance. Unfortunately the network does not converge in that case. Thus, we are not able to show the colour results for the case of luminance-chrominance decomposition. The reason why the network diverge is not known, it may be due to the special statistics of the chrominance information of natural scene.

6. Discussion

In their paper, Doi et al. [8] show that the receptive field of ganglion cells at the output of the retina could be explained as an independent component analysis of natural scene when sampled by the random arrangement of chromatic cones in the retina. Here we show that the spatial and chromatic information could even be reconstructed from the mosaic with a good accuracy. We also show that the approach consisting of estimating luminance gives better results, although we are not able to reconstruct the chrominance by that way.

In this paper we used a supervised implementation of the neural network, given the input and the output to reach the convergence of the network. In the visual system if this kind of reconstruction happens, it remains to explain how the network would converge without having recourse to a supervised method for learning.

References

1. Roorda and D.R. Williams *Nature* **397** (11), pp.520-522 (1999).
2. B.K. Gunturk, J. Glotzbach, Y. Altunbasak, R.W. Schafer, and R.M. Mersereau, *IEEE Signal Processing Magazine*, **22** (1), pp.44-54, (2005).
3. F.A.A. Kingdom, and K.T. Mullen, *Spatial Vision*, **9**, pp.191-219 (1995).
4. E. Hering, «Zur Lehre vom Lichtsinn», Wien 1878.
5. Chaix de Lavarène, D. Alleysson and J. Hérault, *Computer Vision and Image Understanding*, **107**, (1), pp.3-13 (2007).
6. Alleysson, S. Süsstrunk and J. Hérault, *IEEE Trans. Image Processing*, **14** (4), pp. 439 - 449, (2005).
7. Rohde, Lens (Light Efficient Network Simulator) http://tedlab.mit.edu/~dr/Lens/
8. Doi, T. Inui, T-W Lee, T. Wachtler, and T.J. Sejnowski TJ *Neural Computation* **15**(2) pp.397-417, (2003).

THE CONNECTIVITY AND PERFORMANCE OF SMALL-WORLD AND MODULAR ASSOCIATIVE MEMORY MODELS

WEILIANG CHEN,
ROD ADAMS, LEE CALCRAFT, VOLKER STEUBER AND NEIL DAVEY
Science and Technology Research Institute, University of Hertfordshire, College Lane Hatfield, Herts, AL10 9AB, United Kingdom

The mammalian cerebral cortex handling associative memory has a sparse connectivity. In this paper we investigate the effects of network connectivity on the associative memory performance. Two specific models are studied including the *Watt-Strogatz small-world network* and the modular network. The *global and local efficiency* of the connectivity as well as the *Effective Capacity* of associative memory performance are reported. The results are surprising. The best *EC* performance is achieved at the middle level of rewiring in both networks. No significant relationship is found between the *global efficiency* and *EC*, whilst there is a possibly inverse correlation between the *local efficiency* and *EC*.

1. Introduction

The canonical Hopfield Net [1] is usually fully connected. Such networks are tractable for mathematical analysis. However, this connectivity lacks biological plausibility. In a complex neural system like the mammalian cerebral cortex, each neuron is connected to several thousands of other neurons, whilst the total number of neurons can be up to 10^9 (mouse cortex) or 10^{11} (human cortex) [2]. Therefore the mammalian cerebral cortex, the system handling the function of associative memory, is very sparsely connected.

There are two general methodologies to investigate the connectivity of such a complex network. Some recent research investigates aspects of high level connectivity in the mammalian cortex using measures from graph theory [3-5]. Results from these investigations show that at this level the network of the mammalian cortex is not a random network. On the other hand, using a bottom-up methodology, some research focuses on the connectivity between neurons. Some modular structures, such as the "minicolumn" and "column", were proposed as the possible building blocks of the cortex [6].

In any very large, physically realised neural network the nature of their interconnections will be critical to the functionality of the system. However, in such systems there are severe physical constraints which restrict the possible configurations. For example, heat must be dissipated, resources must be globally distributed and sufficient space must be available for all the desired connecting fibre [7]. In this paper we try to find out the relationship between the connectivity of a topological network and its associative memory performance. Results presented here can help to elucidate the basic functionality of associative memory.

2. Measures of the Network Connectivity

2.1. *Path Length, Clustering Coefficient, and the Small-world Network*

Watts and Strogatz [5] investigated some real world networks and discovered that these networks were neither completely regular nor completely random. Graph theoretical measures were used to quantify the properties associated with their connectivity. In particular, two measures, the mean *Path Length* (L), and the *Clustering Coefficient* (C), were introduced.

The *Path Length* is the minimum number of arc traversals to get from one node to another. An average over all pairs of vertices is used to produce L(G) for a graph G. Denoting the length of the shortest path for each pair of vertices as d_{ij}, the *Path Length* of a graph G with N vertices is

$$L(G) = \frac{1}{N(N-1)} \sum_{i \neq j \in G} d_{ij}$$

It is notable that for a disconnected graph, L(G) is problematic since d_{ij} of any pair of disconnected vertices is undefined.

The *Clustering Coefficient* C(G) of a directed graph G is defined as follows. Firstly, define C_i, the local clustering coefficient of node i, as

$$C_i = \frac{\text{\# of edges in G}_\text{i}}{\text{maximum possible \# of edges in G}_\text{i}} = \frac{\text{\# of edges in G}_\text{i}}{k_i(k_i - 1)},$$

where G_i is the subgraph of neighbours of i (excluding i itself), and k is the number of neighbours of vertex i. C_i denotes the fraction of every possible edges of G_i which actually exist. The *Clustering Coefficient* of a graph G, C(G), is then defined as the average of C_i over all vertices i of G:

$$C(\text{G}) = \frac{1}{N} \sum_{i \in G} C_i$$

It is found [5] that a lattice (locally connected network, see figure 1) has both high mean *Path Length* and high *Clustering Coefficient*. On the other hand, a random network has both low mean *Path Length* and low *Clustering Coefficient*. Between these two extreme cases there are a large number of networks which have nearly as low a mean *Path Length* as the lattice (the so-called small-world effect), as well as a high *Clustering Coefficient*. This characteristic (low L, high C) turns out to be a common feature in real networks. Examples of such small-world networks are real neural networks (the cat's cerebral cortex, the neural network of *C.elegans*), social networks and the *World Wide Web* [4, 5, 8].

Watts and Strogatz [5] also proposed a method to construct such networks, which they call *small-world networks*. In their model all vertices are first arranged as a one dimensional ring and are connected to their k nearest neighbours. This network will have both a high L and C. By randomly rewiring a proportion of the connections, the *Path Length* of the network drops significantly, whilst the *Clustering Coefficient* remains at a high level. This is the *small-world network*. If this process continues until all connections are randomly rewired, the network will become a random network with both low L and C. Figure 1 shows the construction of a *small-world network*.

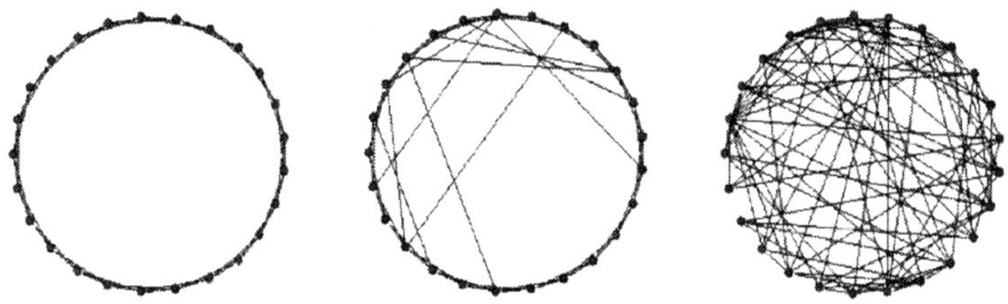

Figure 1. The *W-S* model [5]. Left: A lattice or locally connected network. Middle: A *small-world network* with rewiring probability p = 0.1. Right: A random network (p = 1). In all three cases the number of afferent connections is, k = 4. Diagrams generated with the Pajek package [9]. The left network has both high L and C, whilst the right network has both low L and C. The middle one has low L but high C. (L: mean *Path Length*; C: *Clustering Coefficient*)

2.2. *Global and Local Efficiency of the Network*

Watts and Strogatz [5] characterize the *Path Length* and the *Clustering Coefficient* as two different measures. In fact they can be unified, as shown by Latora and Marchiori [10], to one single measure, the *efficiency* of a network, as well as its subnetworks.

For a directed graph G (connected or disconnected), the average *efficiency* E(G) is defined by the following formula:

$$E(\mathrm{G}) = \frac{1}{N(N-1)} \sum_{i \neq j \in G} \frac{1}{d_{ij}}$$

In particular, the *efficiency* of a fully connected network, which contains all *N(N-1)* edges, is named as $E(\mathrm{G}^{ideal})$. For a topological, directed graph, $E(\mathrm{G}^{ideal})$ = 1. Unlike the mean *Path Length*, E(G) won't be divergent for a disconnected graph because $1/d_{ij}$ is defined as 0 for any disconnected pair of *i,j*.

To formalize the *Path Length* and the *Clustering Coefficient* to a single measure, two new terms, the *global efficiency* and the *local efficiency* are introduced. The *global efficiency* of a graph G, E_{glob}, is defined as

$$E_{glob} = \frac{E(\mathrm{G})}{E(\mathrm{G}^{ideal})}$$

In fact E can be calculated for any subgraph of G. Therefore the local properties of G can be characterized by the *local efficiency*, E_{loc},

$$E_{loc} = 1/N \sum_{i} \frac{E(\mathrm{G}_i)}{E(\mathrm{G}_i^{ideal})}$$

G_i is defined as the subgraph of all the neighbours of vertex *i*. As before G_i^{ideal} is the ideal case of G_i which contains all possible edges. The *small-world network* is now characterized as a set of networks with both high *global* and *local efficiency.*

3. The Connectivity of the Real Mammalian Cortex

Braitenberg and Schüz [2] investigated the connectivity of the mammalian cerebral cortex and suggested a system with two levels of connectivity. At a high level connectivity, the network is constructed mainly from area-to-area excitatory connections between pyramidal cells. At the low level connectivity, the network within an area is constructed from short range excitatory and inhibitory connections of both pyramidal and non-pyramidal cells.

Much research [4, 8, 11] indicates that the area-to-area connectivity has a low *Path Length* but high *Clustering Coefficient* (high *global and local efficiency*), just like a *small-world network* does. On the level of individual neurons, the connectivity is so complex that only some general statistics and hypotheses can be produced [2]. One important hypothesis [6] suggests that the basic functional unit of the mammalian cortex is the "minicolumn", a columnar structure constructed from several hundreds of neurons. Although this hypothesis is still debatable [12], it suggests that the network of an associative memory model could be constructed as a set of interconnected modules.

The existence of two levels of connectivity in the mammalian cerebral cortex lead us to investigate the associative memory performance of two different types of networks, the *Watt-Strogatz (W-S) small-world network* and a modular network, as described in detail later.

4. The High Capacity Associative Memory Model

4.1. *Dynamics*

The units in the network are simple bipolar threshold devices, summing their inputs and firing according to the threshold. The net input, or *local field,* of a unit, is defined by $h_i = \sum_{j \neq i} w_{ij} S_j$, where $S(\pm 1)$ is the current state and w_{ij} is the weight on the connection from unit j to unit i. The update rule of network dynamics is slightly different from the one used in the canonical model

$$S_i' = \begin{cases} 1 & \text{if } h_i > \theta \\ -1 & \text{if } h_i < -\theta \\ S_i & \text{for other cases} \end{cases}$$

where S_i' is the new state of S_i, and θ is the update threshold of the dynamics.

In the traditional model θ is usually set to 0 for simplicity. However, previous experiments indicate that the network performance can be improved using a slightly higher value of θ such as 1 or 2 [13]. The non-zero update threshold also reduces non-convergence of the network by ignoring small changes in the inputs.

Unit states may be updated synchronously or asynchronously. The asynchronous update as well as a symmetric weight matrix guarantees the network will evolve to a fixed point. However, we found that without these restrictions, the network could still achieve fairly similar convergence properties. The synchronous update is suitable for parallel computation, although it increases the chance of non-convergence. Therefore an update threshold of 1 and synchronous update were used in our experiments.

If a trained pattern $\boldsymbol{\xi}^{\mu}$ is one of the fixed points of the network then it is successfully stored and is called a *fundamental memory*.

4.2. *Learning*

A one-shot Hebbian training is commonly used as the standard learning rule of the Hopfield Net. Although simple to implement and also statistically tractable, this learning rule has several drawbacks. The one-shot Hebbian rule does not guarantee that all trained patterns are actually learnt (which means they may not be *fundamental memories*). Furthermore it is widely known that such networks have quite a low theoretical maximum capacity (0.14*N* for a fully connected network with *N* units[14]). The performance of an associative memory can be improved using other classes of learning rules [14]. In our experiments, we adopted and modified Gardner's perceptron learning rule [15] which guarantees all trained patterns with be memorized, as well as given a significantly higher theoretical maximum capacity of up to 2*N* for unbiased patterns. The detailed training process is given as follows:

Denoting T as the learning threshold
Begin with a zero weight matrix
Repeat until all units are correct

Set the state of the network to one of the ξ^p
For each unit, i, in turn:
Calculate its local field h_i^p
If ($\xi_i^p h_i^p < T$) *then change the weight on connections into unit i according to:*

$$\forall i \neq j \quad w'_{ij} = w_{ij} + C_{ij} \frac{\xi_i^p \xi_j^p}{N}$$

where $\{C_{ij}\}$ is the connection matrix

End For
End

4.3. *Performance Measures*

It is important to investigate not only the capacity of the associative memory model but also the ability of *fundamental memories* to act as *attractors* in the state space of the network dynamics.

To measure this we use the *Effective Capacity* of the network, *EC* [16, 17]. The *Effective Capacity* of a network is a measure of the maximum number of patterns that can be stored in the network with *reasonable* pattern correction still taking place. We take a fairly arbitrary definition of *reasonable* as correcting the addition of 60% noise to within an overlap of 95% with the original fundamental memory. Varying these figures gives differing values for *EC* but the values with these settings are robust for comparison purposes (see [17] for the effect on *Effective Capacity* of varying the degree of applied noise, and the required degree of pattern completion). For large fully-connected networks the *EC* value is about 0.1 of the maximum theoretical capacity of the network, but for networks with sparse, structured connectivity *EC* is dependent upon the actual connection matrix *C*.

The *Effective Capacity* of a network is defined as follows:

```
Initialise the number of patterns, P, to 0
Repeat
  Increment P
  Create a training set of P random patterns
  Train the network
  For each pattern in the training set
    Degrade the pattern randomly by adding 60% of noise
        With this noisy pattern as start state, allow the
          network to converge
        Calculate the overlap of the final network state with
          the original pattern
  End For
  Calculate the mean pattern overlap over all final states
Until the mean pattern overlap is less than 95%
The Effective Capacity is P-1
```

4.4. *Examined Connectivity*

We examined two different types of network, the *W-S small-world network* [5] and a modular network.

4.4.1. *The W-S Small-world Network*

We constructed the small-world network according to Watts' and Strogatz's method[5] (See 2.1. for details). All N units are arranged as a one dimensional ring and are locally connected with k ($0 < k < N$) nearest neighbors. After that, for each unit, a proportion of connections are rewired, giving a rewiring rate of p. The rewiring rate p is increased from 0 (which defines a local lattice) to 1 (which defines a random network) by an increment of 0.1. Different N and k were examined, here we present results of two series, one with N = 600, k = 199, and another one with N = 2000, k = 199 (the value 199 was chosen due to the fact that it is the number of connections per unit in a 200 units, fully connected module. See 4.4.2.).

4.4.2. *The Modular Network*

A large scale associative memory network can also be constructed from discrete networks, which we refer to as *modules*. In our implementation, each module is a fully connected network. To construct a whole network from these modules, a proportion (defined by q) of intra-modular incoming connections of each unit were rewired as inter-modular connections from units in other modules. See Figure 2 for more details.

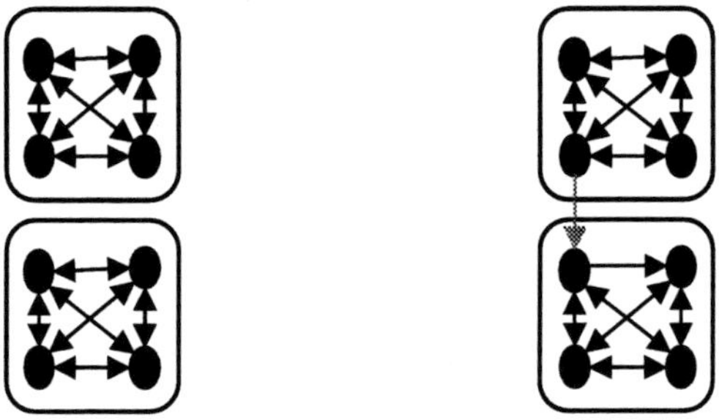

Figure 2. The modular network. Left: The initial structure of two discrete *modules*, each of them is a fully connected network. Right: The *modules* are connected by rewiring an incoming connection of a unit is rewired as inter-modular connection from a unit in another *module* (the red one). The rewiring will take place in selected connections of all units.

In our experiments, the number of units in a module was always set to 200, therefore the number of connections per unit, k, is 199. We examined two modular networks with different numbers of modules. One has 3 modules (N = 600) and the other has 10 modules (N = 2000). The rewiring proportion q was increased from 0 (as discrete modules) to 1 by an increment of 0.1.

5. The Results

All experiments were repeated 10 times and the results are averaged. The 95% confidence intervals are given. Here we report the results of *global and local efficiency*, as well as the *Effective Capacity* performance of the networks.

5.1. *The W-S Small-world Network*

Figures 3 and 4 give the main results for the W-S networks (Figure 3: N = 600; Figure 4: N = 2000). For both networks, the *global efficiency* of the networks increases abruptly. However, it becomes stable as p changes from 0.1 to 1, which suggests that the *global efficiency* may have little relationship with the performance of associative memory models. On the other hand the *local efficiency* of the network decreases rapidly at the start and then changes slowly.

It is noticeable that the *Effective Capacity* of the network increases quickly at the beginning and saturates later. As described in Figure 1, a rewiring rate p = 0 indicates a local connected network, p = 1 indicates a random network. The saturation of *EC* first happens at a mid-value and remains when the network becomes more and more random as p moves towards 1. This phenomenon suggests that a *small-world network* is a better choice than the random network because it can achieve the same associative memory performance with less cost of wiring.

Another thing worth mentioning on these figures is the possible inverse correlation between *EC* and the *local efficiency*. Both of these measures seem to change rapidly at beginning and saturate at a similar rewiring rate later. More investigations on this relationship are ongoing.

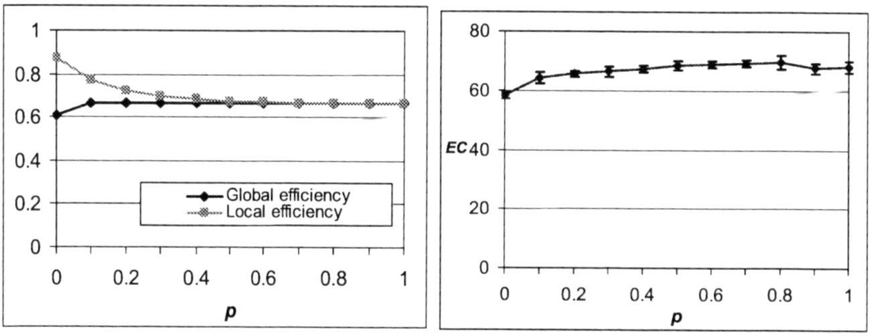

Figure 3. Results for the *W-S* network (N = 600, k = 199). Left: Normalized *Global and local efficiency* of the network. Right: *Effective Capacity* of the network.

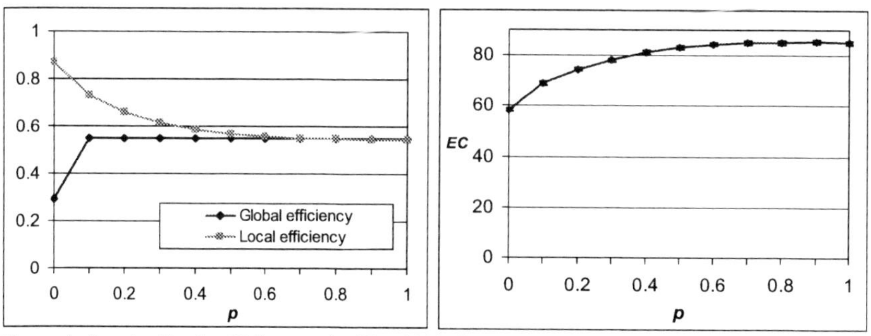

Figure 4. Results for the *W-S* network (N = 2000, k = 199). Left: Normalized *Global and local efficiency* of the network. Right: *Effective Capacity* of the network.

5.2. *The Modular Network*

Figures 5 and 6 report the results for the modular networks (Figure 5: N = 600, 3 *modules*; Figure 6: N = 2000, 10 *modules*). The most significant difference between the *W-S* network and the modular network is the results prior to rewiring. The *EC* prior to rewiring for the modular network is a lot lower than for the *W-S* network. This is due to the fact that the modules prior to rewiring are disconnected from each other and act individually as a 200 units, fully connected associative memories. Note that a set of fully connected networks will always have a *local efficiency* of 1, whilst the *W-S* network starts from approximately 0.9. The *EC* of the modular network rapidly increases and achieves similar result to the one of *W-S* network at a low q (about 0.3). Similar properties of the *W-S* network, such as the early saturation of *EC* and *efficiency* of non-complete random networks, and the possible inverse correlation of *EC* and *local efficiency*, were also found on the modular network.

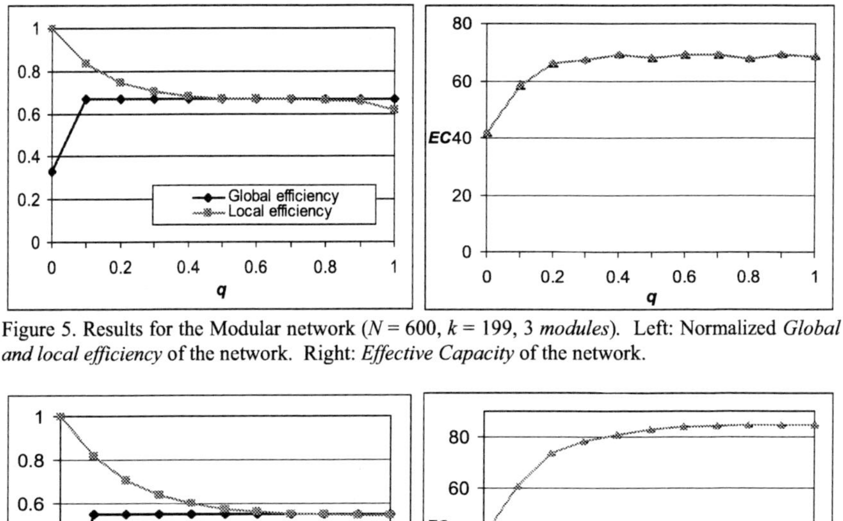

Figure 5. Results for the Modular network ($N = 600$, $k = 199$, 3 *modules*). Left: Normalized *Global and local efficiency* of the network. Right: *Effective Capacity* of the network.

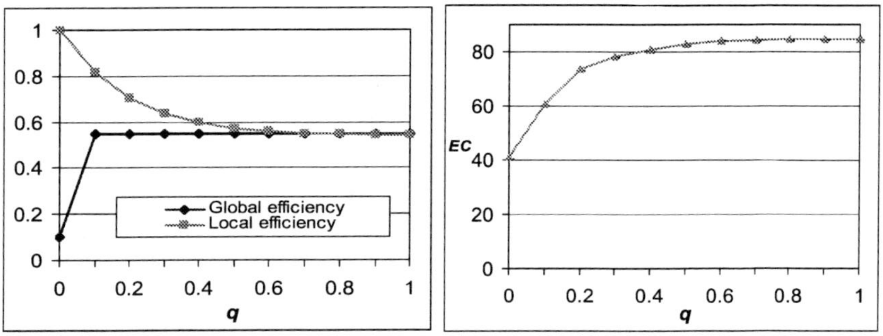

Figure 6. Results for the Modular network ($N = 2000$, $k = 199$, 10 *modules*). Left: Normalized *Global and local efficiency* of the network. Right: *Effective Capacity* of the network.

6. Conclusion and Discussion

In this paper we investigated how the connectivity of sparse associative memory models affects network performance. Two different types of network were examined, the *Watts-Strogatz small-world* network and a modular network.

Global and local efficiency were used to characterize the network connectivity and *Effective Capacity* was used to measure the networks' performance as an associative memory.

Our main result suggests that rewiring in both models improves the network's performance as an associative memory. The best performance is achieved at about p (or q) = 0.5. The *global efficiency* saturates quickly in all experiments (p or q = 0.1) whilst the *local efficiency* decreases to roughly the point where EC is maximum. It is interesting that the EC of a modular network with only intra-modular connections is very poor. However, the performance can be considerably improved by introducing inter-modular connections. In fact, the EC performance of both networks is similar for all rewiring rates higher than 0.3.

From the results there seems to be a roughly inverse correlation between the values of *local efficiency* and EC. However, this relationship is still far from clear and currently under further investigation.

This paper focuses on the topological properties of the network connectivity. However, it is also important to investigate the connectivity of networks where a distance metric is defined. Details of these investigations can be found in our recent papers [16, 18].

References

1. Hopfield, J.J., *Neural networks and physical systems with emergent collective computational abilities.* Proceedings of the National Academy of Sciences of the United States of America - Biological Sciences, 1982. **79**: p. 2554-2558.
2. Braitenberg, V. and A. Schüz, *Cortex: Statistics and Geometry of Neuronal Connectivity.* 1998, Berlin: Springer-Verlag.
3. Sporns, O., *Graph theory methods for the analysis of neural connectivity patterns*, in *Neuroscience Databases. A Practical Guide*, R. Kötter, Editor. 2002: Klüwer, Boston, MA. p. 171-186.
4. Sporns, O. and J.D. Zwi, *The small world of the cerebral cortex.* Neuroinformatics, 2004. **2**(2): p. 145-62.
5. Watts, D.J. and S.H. Strogatz, *Collective dynamics of 'small-world' networks.* Nature, 1998. **393**(6684): p. 409-10.
6. Mountcastle, V.B., *The columnar organization of the neocortex.* Brain, 1997. **120**(4): p. 701-722.
7. Markram, H., et al., *Interneurons of the neocortical inhibitory system.* Nature Reviews Neuroscience, 2004. **5**(10): p. 793-807.
8. Milgram, S., *The Small World Problem.* Psychology Today, 1967: p. 60-67.

9. de Nooy, W., A. Mrvar, and V. Batagelj, *Exploratory Social Network Analysis with Pajek*, in *Structural Analysis in the Social Sciences*. 2005, Cambridge University Press.
10. Latora, V. and M. Marchiori, *Economic small-world behavior in weighted networks.* The European Physical Journal B-Condensed Matter, 2003. **32**(2): p. 249-263.
11. Watts, D. and S. Strogatz, *Collective Dynamics of 'small-world' networks.* Nature, 1998. **393**: p. 440-442.
12. Horton, J. and D. Adams, *The cortical column: a structure without a function.* Philosophical Transactions of the Royal Society B: Biological Sciences, 2005. **360**(1456): p. 837.
13. Chen, W., et al., *The Effects of Changing the Update Threshold in High Capacity Associative Memory Model.* Proceedings of the Sixth International Conference on Recent Advances in Soft Computing (RASC2006), 2006.
14. Abbott, L.F., *Learning in neural network memories.* Network: Computational Neural Systems, 1990. **1**: p. 105-122.
15. Gardner, E., *The space of interactions in neural network models.* Journal of Physics A, 1988. **21**: p. 257-270.
16. Calcraft, L., R. Adams, and N. Davey, *Locally-Connected and Small-World Associative Memories in Large Networks.* Neural Information Processing - Letters and Reviews, 2006. **10**(2): p. 19-26.
17. Calcraft, L., *Measuring the Performance of Associative Memories*, in *Computer Science Technical Report.* 2005, University of Hertfordshire.
18. Calcraft, L., R. Adams, and N. Davey, *Efficient architectures for sparsely-connected high capacity associative memory models.* Connection Science, June 2007. **19**(2), p. 163-175.

CONNECTIONIST HYPOTHESIS ABOUT AN ONTOGENETIC DEVELOPMENT OF CONCEPTUALLY-DRIVEN CORTICAL ANISOTROPY

MARTIAL MERMILLOD[A], NATHALIE GUYADER[B], DAVID ALLEYSSON[C], SERBAN MUSCA[D], JULIEN BARRA[C] and CHRISTIAN MARENDAZ[C]

[A]LAPSCO CNRS UMR 6024, Clermont-Ferrand, [B]GIPSA-lab CNRS UMR 5216, Grenoble, [C]LPNC CNRS UMR 5105, Grenoble, [D]Universidad deDeusto, Bilbao.

The physiological structure of the striate cortex is composed of neural columns sensitive to different orientations [9] with an over-representation of cardinal orientations [13]. As an explanation, it seems that statistics of the power spectra of natural images have shown that vertical and horizontal orientations are most present in our visual environment [3, 12, 20]. The over-representation of horizontal and vertical orientation columns in visual cortex was interpreted as an adaptation of the visual system to this greater amount of horizontal and vertical information. The present study is an attempt to confirm that the information about these two orientations is quantitatively more available in the environment, but also that cardinal information is qualitatively more efficient than other orientations for a connectionist model of categorisation.

1. Introduction

Primary visual cortex neurophysiology shows a columnar organization that is differentially sensitive to spatial frequency bands [18]. In addition, specific columns of cells code the different orientations of the visual environment [1, 9]. Nevertheless, all the orientations are not equally represented at a physiological level. Striate cortex provides more cells for vertical and horizontal orientations, generating an over-representation of these orientations in the human visual system [1, 2, 13].

Over the last two decades, a number of researchers have shown that correct processing of visual information requires more than a good understanding of the neurophysiological structure of V1. Since Field [4], research in visual cognition is taking into account the fact that the analysis of the statistics in the visual environment can lead to a better understanding of neurophysiological data. In other words, the statistics of natural environments are, in fact, necessary to understand the internal processes occurring in the cortical areas to build up an efficient representation of the environment [4, 5]. In particular, Van der Schaaf & Van Hateren [20] have shown that natural images contain more information in the horizontal and vertical orientations. At a computational level, that view is supported by results showing that V1 receptive fields seem to evolve as an adaptation to the visual environment [14].

Moreover, Dragoi, Turcu & Sur [3] have shown that cortical responses to natural scenes are more narrowly tuned to specific orientations, with more cardinally tuned (orientated) cells than other types of cells. The result was a greater firing-rate stability of neurons coding cardinal orientations (vertical and horizontal) than neurons coding oblique orientations. Hence, we decided to test the assumption that the different orientations of our visual environment do not have the same usefulness for categorization, which is one of the most important cognitive functions. Specifically, our goal was to test the categorization properties of one specific orientation channel independently of the amount of data provided by that specific channel. This assumption follows from the approach proposed by Field (1987). We suppose that the V1 physiological structure depends on visual environment statistics. In addition, we suppose that categorization processes could be improved by the use of horizontal and vertical orientation information, independently of the amount of energy provided by the different orientation channels. In other words, we suppose that cardinal orientations, independently of the amount of data provided to the cognitive system (because data will be normalized to be equivalent for each channel), will be more useful to categorize natural images. In a first simulation, we tested this hypothesis on simple visual objects without background. In a second simulation, we tested the same hypothesis on a broader pattern of natural scene images.

2. Simulation of the primary visual system

For the simulation, each picture was converted from color to grey level varying from 0 to 255 and was resized in a 256*256 picture matrix. Each image was then processed by the perceptual model. This model simulated the retina pre-processing [8]. The retinal photoreceptors do a space-time high-pass filtering after an adaptive compression process. This results in a contrast equalization of the image, hence a relative insensitivity to illumination variations, and a spectral whitening which compensates for the 1/f amplitude spectrum.

After the retinal pre-processing a Hanning window was applied to avoid a boundary effect (i.e. an over-representation of the vertical and horizontal orientations in the Fourier domain). Images were then transferred in the Fourier domain and filtered by a set of Gabor filters, simulating cortical cells. Gabor functions were used because they "fit the 2D spatial and spectral structure of simple cells in visual primary cortex, with a small non-structured error indiscernible from random error" [10, 11]. Each filter corresponds to one spatial frequency and one orientation. The filtered outputs were used to determine energy coefficients by coding the local energy spectra, which is the square of the amplitude spectra. Energy coefficients were used to simulate V1 complex cells [17]. Each step of our model is showed in Figure 1.

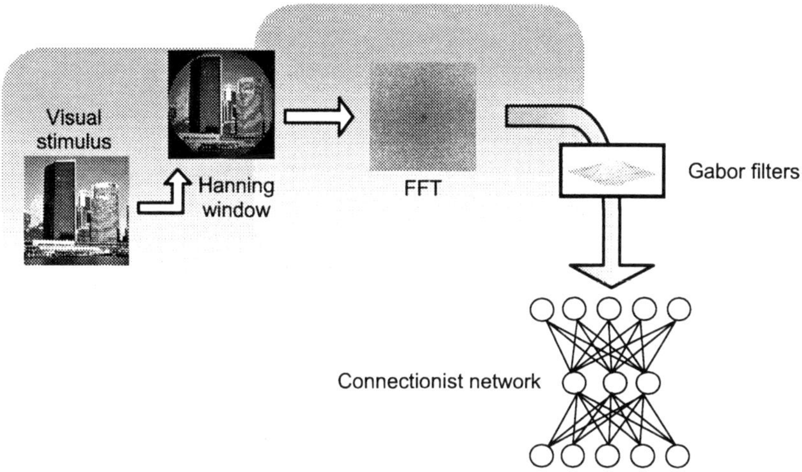

Figure 1. Stimuli pre-processing and the connectionist network.

In keeping with biological data, we used a relative radial width of 1 octave for each spatial frequency channels [1]. The orientations of the filters were set to 0°, 30°, 60°, 90°, 120° and 150° for each of the 5 spatial frequency band. Each filter provides a mean energy value; thus, our model describes each image with a vector of length 30.

3. First connectionist simulation: natural objects

3.1. *Hypothesis*

The goal of this simulation is to test categorization performance depending on different orientations with a non-linear connectionist model. Pilot simulations based on a standard back-propagation neural network showed that the hetero-association process often reached a ceiling effect among the different categories that does not allow us to conclude a clear advantage for one or another orientation channel. In order to overcome that over-learning of the network, we decided to test another training algorithm by using an auto-associator as neural network classifier. The model was designed to create an "internal representation" at the hidden layer level and to test this representation with new exemplars from the same or from other categories.

This method was used on a similar categorization task to simulate human behavior [6, 7]. That method was called "self-supervised" neural network and was used to simulate categorization in a more ecological manner than a completely supervised hetero-associator. The training algorithm operates as follows. First, a stimulus (i.e., an energy vector produced by our perceptual model) is presented to the input layer of the network; the resulting activation is propagated through the neural network and produces an output activation; then, this output activation is associated with the theoretical correct vector (the same as the input vector) and error is back-propagated to reduce the discrepancy between the observed output vector and the theoretical output expected; and then a new vector from the same category is auto-associated with itself, and so forth. The aim of the algorithm is to create at the hidden layer level an "internal representation" of one specific category at a time. Therefore, if new exemplars from the *same category* are correctly classified by the autoencoder, we should observe a *low error* (in terms of Euclidean distance) increase between the observed output produced by the network and the desired one (i.e. the input). If exemplars from a *novel category* are recognized as different from the previously learned category, we should observe a significant *increase of error*. We expect a significantly larger increase of error with new category exemplars when the model was trained on 0° and 90° Gabor filters than when the network was trained on the other tuned (orientated) filters.

3.2. *Material and method*

The first stimuli were used in a developmental psychology categorization task during the last decade [16]. These visual images were demonstrated to contain sufficient information to be categorized by the early human perceptual system (3- and 4-month-olds infants). Stimuli consist of a set of 70 images classifiable in 5 basic-level categories: "human", "horse", "cat", "fish" and "car" (Figure 2). Those categories consisted of 18 human exemplars, 18 horse exemplars and 18 cat exemplars. Additionally, there were 8 fish and 8 cars only used as test categories.

Figure 2. Example of pictures used in the first simulation. Pictures belong to five categories: "human", "horse", "cat", "fish" and "car".

Each Gabor filter produces an average energy value depending on the amplitude spectrum at its location. Thus, for each filter, we have 70 responses corresponding to the 70 images filtered by the model. We used as inputs the 5 energy values provided by the filters corresponding to the 5 spatial frequency bands on a specific orientation (0°; 30°; 60°; 90°; 120° and 150°). Data were normalized between 0 and 1 across the 5 categories (in order to keep the distribution property of each filter) and independently for each specific orientated filter. In other words, the categorization process done by the neural network cannot use the amount of energy from a specific orientation to categorize the different exemplars because the range of energy for each specific orientation was normalized to be equivalent.

The length-5 vectors were auto-associated by a 3-layer back-propagation network. The network architecture consists of 5 input units, 5 output units and 3 hidden units (Figure 1). The learning rate was fixed to 0.1, momentum to 0.9 and the Fahlman offset to 0.1. We generated internal representation by training the network on 17 exemplars of the training category. The training consists of associating each exemplar with itself for 500 epochs. The neural network was trained on one category at a time, and then tested on the remaining exemplar from the training category Vs one exemplar from the four other categories. Results were averaged over 50 runs of the above training-test procedure. The measure was the mean Euclidian distance between the output produced by the tested vector and the desired output.

3.3. *Results*

The first result (Figure 3) is that there was an effect of better convergence with cardinal orientations, especially 90° filters (coding for horizontal orientations). The convergence of the network was defined as the ability of the network to learn a training category by decreasing the error between the expected output and the observed output produced by the network. A post-hoc Scheffé on all orientations reveals that only 90° filters produced a better convergence than all other filters (p<.001). The second best filter is the 0° filter. However, the Scheffé test reveals that this result is not significantly better than oblique orientations (in particular, 60°. filters and 120°. filters).

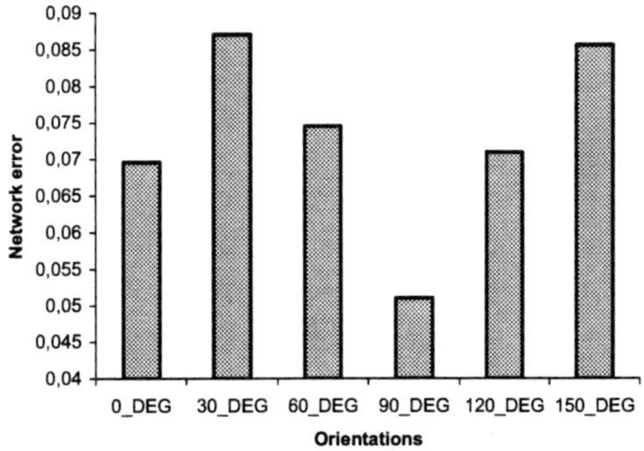

Figure 3. Average errors produced by the neural network for training item categorization depending on the different oriented filters.

At the increased level of error for new category exemplars (Figure 4), 90° filters produced the expected significant increase of error for new categories compared to the increase of error produced by data from 0° filter data ($F(1, 1782) = 165.6$, $p < .001$), 30° filters ($F(1, 1782) = 438.29$, $p < .001$), 60° filters ($F(1, 1782) = 237.73$, $p < .001$), 120° filters ($F(1, 1782) = 315.21$, $p < .001$) and 150° filters ($F(1, 1782) = 673.13$, $p < .001$). A Scheffé post-hoc demonstrates that the differences remain after a post-hoc correction (p<.001), but also that 0° filters produced a significantly better increase of error than all other orientations (p<.001) except 60° filters (p>.05).

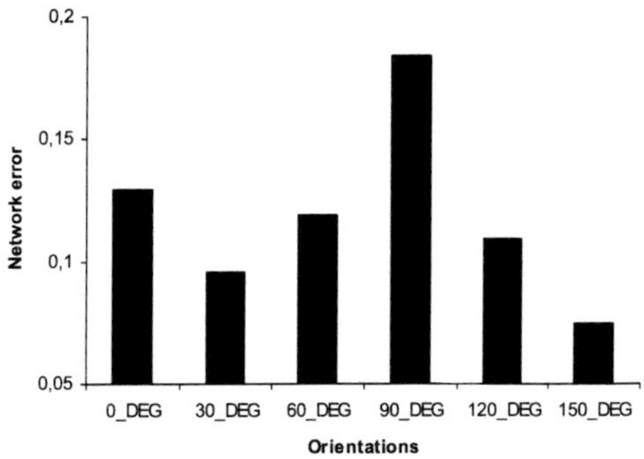

Figure 4. Average difference of error between new categories and training category for different filter orientations.

3.4. *Discussion*

When the neural network was trained on the energy values provided by horizontal information, it produced a more reliable internal representation and an improved ability to differentiate the different categories. It is important to note that the improvement was obtained after data normalisation across the different orientations. This means that we made it impossible for the neural network to use the quantity of information across different orientation channels to differentiate the categories. The information provided by horizontal Gabor filters (i.e. the vertical orientations) produced better categorization efficiency than certain other oblique orientations, but not all. Moreover, the performance of the 0° filters was significantly below that of 90° filters. However, the effect was tested on natural objects extracted from their background, the effect could be generated by the specificity of the categories or the absence of contextual information. Thus, we decided to test the orientation effect on a broader set of more complex exemplars from natural scene images.

4. Second connectionist simulation: natural scenes images

4.1. *Hypothesis*

The goal of this simulation is to test the categorization performance of a connectionist model depending on the orientations of filters. We expected a significantly larger increase of error with new category exemplars when the model was trained on 0 and 90° Gabor filters than when the network was trained on the other tuned or orientated filters.

4.2. *Material and method*

The simulation was the same as above: images were decomposed by 30 Gabor wavelets into 5 spatial frequency bands and 6 different orientations. The stimuli used correspond to 6 categories of natural scenes of 12 stimuli each: Beach, City, Forest, Mountain, Indoor Scene and Village (Figure 5). Each Gabor filter produced an average energy value depending on the amplitude spectrum under its location. Thus, for each filter, we have 72 responses corresponding to the 72 images filtered by the model.

The neural network was trained on 6 patterns of data, corresponding to the 6 orientations of the bank of filters (0° – 30° – 60° – 90° – 120° and 150°) for each category. We used as inputs the 5 averaged energy values provided by the 5 frequency tuned filters for a specific orientation. Data were normalized between 0 and 1 for each filter independently. The neural network architecture

Figure 5. Category exemplars used in the second simulation: City, Indoor, Forest, Mountain, Beach and Village.

and parameters were kept identical compared to the previous simulation. The network was trained independently on 11 exemplars of each category and then it was tested on the remaining exemplar of the training category vs. one exemplar from the other categories. Results were averaged over 50 runs of the above training-test procedure on each training category. The measure was the averaged Euclidian distance between the output produced by the tested vector and the desired output.

4.3. *Results and discussion*

The neural network reached a reliable level of convergence with each individual orientation (Figure 6). Nevertheless, 0° and 90° oriented filters produced a significantly lower error rate than other orientations (average error = .098 for 0° and .103 for 90° filters, $\underline{F}(1, 1764) = 613.28$, $\underline{p} < .001$). A Scheffé post-hoc revealed that the errors for the two cardinal orientations were not significantly different, but were both significantly below all other oblique orientations.

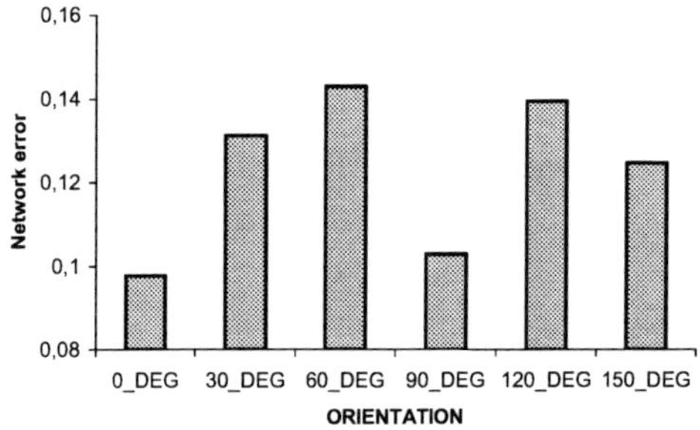

Figure 6. Average errors produced by the neural network when tested on new exemplars from the training category depending on the different oriented filters.

As before, the second result is not a measure of convergence on the learning category but a measure of the error increase when the network is trained on a new category. As expected, 0° filters produced a significantly greater increase of error (average increase of error: .187). A Scheffé post-hoc demonstrates that only 0° filters categorized significantly better than each of the other oriented filters ($p<.001$). In other words, the horizontal filters coding vertical orientations produced a better increase of error than other orientations (Figure 7). However, the unexpected result was the small increase of error produced by the 90° oriented filters which did not produce the expected data (average error: .143). Despite a better convergence for the two types of tuned filters — meaning an ability to learn a specific category with 0° and 90° filters — only 0° filters produced a better increase of error when tested on new category exemplars.

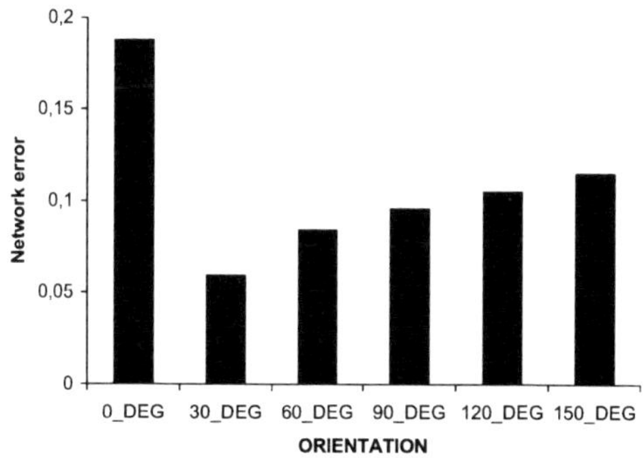

Figure 7. Average difference of error between exemplars from new categories and training category for different oriented filters.

5. Conclusion

In an attempt to test the efficiency of the striate cortex organization of oriented cell columns, we have shown that vertical and horizontal orientations provided by a simulation of the human perceptual system seem to be more efficient for a distributed classifier in a perceptual categorization task. Specifically, an advantage was shown for horizontal and vertical orientations in a visual object categorization task, whereas only vertical orientation seems to be more useful in natural scene categorization task.

Based on seminal studies on real-world statistics [12, 20] and their relationship to previous neurophysiological data [1, 13] to more recent studies trying to use those statistics to perform classification and identification tasks [15, 19], a link is emerging between the structure of the natural environment, perceptual cortex neurophysiology and high level cognition. However, visual categorization and identification was so far achieved *using* real world statistics as a way to describe specific categories. In recent studies by Torralba & Oliva [19], the shape of the power spectra was used to classify natural images depending on the distribution of energy across different characteristic frequency channels and orientations. Their results showed that super-ordinate categorization, such as man-made scenes and natural scenes, can be differentiated on the basis of their spectral structure. Man-made scenes have a spectrum profile largely dominated by the cardinal orientations whereas natural scenes are much more isotropic across all orientations. Classification tasks on those spectral signatures with PCA or discriminant analysis allow reliable categorization performance at different semantic levels (from super-ordinate to basic level categorization task).

Our study was very different but complementary to that approach and consisted of testing independently each cortical orientation channel in a connectionist classifier. Instead of using the distribution of each category along the orientation channels, we made those distributions equivalent by data normalisation in order to test the efficiency of each individual orientation channel independently of the amount of information provided by each one.

By taking into account the striate cortex cortical cell distribution, the over-representation of cardinal orientation could provide the cognitive system with an advantage in classifying information from the visual environment. In other words, the over-representation of cardinal orientations in the perceptual system could be interpreted i) at a quantitative level, as an adaptation to the amount of information contained in the natural environment ii) at a qualitative level, as an efficient way to encode a more stable representation of different stimuli [3], thereby providing more reliable information to perform visual categorization. Finally, we did not compare the effect of the oriented cell columns on other high level cognition tasks, such as motion perception or visual recognition. Further advantages for those orientations could emerge in cognitive tasks other than visual categorization.

6. Acknowledgments

This work has been supported by a post-doctoral grant from the Fyssen foundation, the National Center for Scientific Research (CNRS UMR 6024) in addition to a grant from the French National Research Agency (ANR Grant BLAN06-2_145908) to MM.

7. References

1. De Valois, R.L., De Valois K.K. (1988). Spatial Vision. *Oxford university press*. New York.
2. De Valois, R.L., Yund E.W. & Hepler, N. (1982). The orientation and direction selectivity of cells in macaque visual cortex. *Vision Research, 22(5)*. 531-544.
3. Dragoi V., Turcu C.M., Sur M. (2001). Stability of cortical responses and the statistics of natural scenes. *Neuron, 32*, 1181-1192.
4. Field, D.J. (1987). Relations between the statistics of natural images and the response properties of cortical cells. *Journal of Optical Society of America, 4,* 2379-2394.
5. Field, D.J. & Brady N. (1997). Visual sensitivity, blur and the sources of variability in the amplitude spectra of natural scenes. *Vision Research, 37(23)*. 3367-3383.
6. French, R. M., Mareschal, D., Mermillod, M. & Quinn, P. C. (2004). The Role of Bottom-up Processing in Perceptual Categorization by 3- to 4-month-old Infants: Simulations and Data. *Journal of Experimental Psychology : General, 133(3),* 382-397.
7. French R. M., Mermillod M., Quinn P. C. & Mareschal D. (2001). Reversing category exclusivities in infant perceptual categorization: simulations and data. Proceedings of the 23th Annual Cognitive Science Society Conference, LEA, 307-312.
8. Guyader, N. Hérault J. (2001). Représentation espace-fréquence pour la catégorisation d'images. in GRETSI. Toulouse.
9. Hubel, D.H., Wiesel, T.N. & Stryker, M.P. (1977). Orientation columns in macaque monkey visual cortex demonstrated by the 2-deoxyglucose autoradiographic technique. *Nature 269(5626)*. 328-330.
10. Jones, J.P. & Palmer L.A. (1987). The two-dimensional spatial structure of simple receptive fields in cat striate cortex. *Journal of Neurophysiology, 58(6),* 1187-1211.
11. Jones, J.P., Stepnoski A. & Palmer L.A. (1987). The two-dimensional spectral structure of simple receptive fields in cat striate cortex. *Journal of Neurophysiology, 58(6),* 1212-1232.
12. Keil, M.S., & Cristobal G. (2000). Separating the chaff from the wheat: possible origins of the oblique effect. *Journal of Optical Society of America, 17(4)*, 697-710.
13. Mansfield R.J. (1974). Neural basis of orientation perception in primate vision. *Science. 186(4169)*. 1133-1135.
14. Mayer N., Herrmann J.M., & Geisel T. (2001). Signatures of natural image statistics in cortical simple cell receptive fields. *Neurocomputing, 38(40)*, 279-284.

15. Oliva, A. & Torralba, A. (2001). Modeling the Shape of the Scene: A Holistic Representation of the Spatial Envelope. *International Journal of Computer Vision, 42(3),* 145-175.
16. Quinn, P. & Eimas, P. (1998). Evidence for a global categorical representation of humans by young infants. *Journal of Experimental Child Psychology, 69,* 151-174.
17. Sakai K. & Tanaka S. (2000). Spatial Pooling in the Second-order Spatial Structure of Cortical Complex Cells. *Vision Research, 40,* 855-871.
18. Tootell, R.B., Silverman M.S. & De Valois, R.L. (1981). Spatial frequency columns in primary visual cortex. *Science, 214 (13).* 813-815.
19. Torralba, A., Oliva, A. (2003). Statistics of Natural Images Categories. *Network: Computation in Neural Systems, 14,* 391-412
20. Van der Schaaf, A., & Van Hateren, J. H. (1996). Modelling the power spectra of natural images: statistics and information. *Vision Research, 36(17),* 2759-2770.

AUTHOR INDEX